POLICY NETWORKS UNDER PRESSURE

For my daughter Line and my son Esben

Policy Networks under Pressure

Pollution Control, Policy Reform and the Power of Farmers

CARSTEN DAUGBJERG
Department of Cooperative and Agricultural Research
South Jutland University Centre
Denmark

Routledge
Taylor & Francis Group

LONDON AND NEW YORK

First published 1998 by Ashgate Publishing

Reissued 2018 by Routledge
2 Park Square, Milton Park, Abingdon, Oxon, OX14 4RN
52 Vanderbilt Avenue, New York, NY 10017

Routledge is an imprint of the Taylor & Francis Group, an informa business

Publisher's Note
The publisher has gone to great lengths to ensure the quality of this reprint but points out that some imperfections in the original copies may be apparent.

Disclaimer
The publisher has made every effort to trace copyright holders and welcomes correspondence from those they have been unable to contact.

A Library of Congress record exists under LC control number: 97076942

ISBN 13: 978-1-138-36294-9 (hbk)
ISBN 13: 978-1-138-36297-0 (pbk)
ISBN 13: 978-0-429-43183-8 (ebk)

Contents

Tables

Acknowledgements

Many people have been involved in the preparation of this book. Some of them helped to develop my ideas without falling into the many pitfalls of political science. Others assisted with the English language, gave me interviews or assisted in collecting data. I very much appreciate this support. The funding of the Nordic Council of Ministers, the Danish Strategic Environmental Research Programme, the Nordic Academy for Advanced Studies, the Danish Research Academy and Aarhus University Research Foundation enabled me to write the book. I am grateful for their funding. The first version was written while I was working in the Department of Political Science and in the Centre for Social Science Research on the Environment, Aarhus University. It was completed in the Department of Cooperative and Agricultural Research, South Jutland University Centre. I would like to thank the people affiliated to these three research institutions for various sorts of help.

Several people have read earlier versions of this book. They deserve separate thanks. Peter Munk Christiansen (Aarhus University) read many drafts during the research process and gave me valuable comments which helped make my points clearer and more convincing. Erik Damgaard (Aarhus University), Lennart J. Lundqvist (Gothenburg University), Jacob Buksti (Copenhagen), Johannes Michelsen (South Jutland University Centre), Arild Vatn (The Agricultural University of Norway), Jørgen Grønnegaard Christensen, Peter Nannested (both at Aarhus University), Jørgen Goul Andersen (Aalborg University) and Hanne Foss Hansen (Copenhagen University) also read an earlier version and pointed out where it could be improved.

I owe a great debt to David Marsh (Strathclyde University, now at Birmingham University) for outstanding hospitality, for giving me constructive criticism and advice, for taking the time for inspiring discussions and for encouraging my work. He also taught me a lot about the English language.

A special vote of thanks also goes to the many others whose advice during the research process was of great value. They read draft chapters or one or more of the five working papers (Daugbjerg, 1993, 1994a, 1995a, 1995b, 1996a) of the research project or assisted in other ways. Wyn Grant (University of Warwick) commented on several parts of my work and shared his insights on agricultural politics and policy with me. Asbjørn Nørgaard (Aarhus University) was always willing to comment on my work and help resolve some of its problems. Søren Winter and Niels Christian Sidenius (both at Aarhus University) read critically several drafts and always had well-considered and detailed comments. I am also thankful to Eugene Bardach (University of California at Berkeley), Inge Bertelsen (Skejby), Elizabeth Bomberg (University of Stirling), Göran Bostedt (Sundsvall University), Jens Blom-Hansen (Aarhus University), Jack Brand (Strathclyde University), Anders Broo (Örebro University), Mick Cavanagh (Strathclyde University), Peter Dobers (Gothenburg School of Economics and Commercial Law), Svante Forsberg, (Gothenburg University), Søren Frandsen (The Institute of Agricultural Economics), Michael Huber (Aarhus University, now at Hamburg University), Alf Inge Jansen (University of Bergen), Flemming Just (South Jutland University Centre), Claire Reid Mackie (Strathclyde University), Peter May (University of Washington, Seattle), Martin J. Smith (Sheffield University), Gert Tinggaard Svendsen (The Aarhus School of Business), Jens Peter Frølund Thomsen (Aarhus University), Marianne Lybek Witt (Hadsten) and the participants in the conference workshops/panels and colloquiums at Strathclyde University. I also benefited from the views of the students attending the course on policy networks and public policy in the Department of Political Science at Aarhus University

Mary B. Muir (East Kilbride, Scotland) put great effort into reading the final draft manuscript and improving my English. She taught me a lot about the language so that hopefully I can do it better next time. She deserves wholehearted thanks for her assistance. Annette B. Andersen and Lisbeth Widahl (both of Aarhus University) deserve special thanks for assisting with the language and many other tasks. So does Mette Sprogøe Guldberg (South Jutland University Centre) for doing the final proof reading and the lay out. I am of course responsible for any errors which may still be there.

Most of the theoretical chapters were written in draft form during my eight months stay in the Department of Government at the University of Strathclyde, Glasgow. I am grateful to David Marsh in particular but also staff, PhD students and secretaries for providing the very best conditions for a pleasant and rewarding stay. The research project also included stays in the Department of Political Science, Gothenburg University. I would particularly like to thank Lennart J. Lundqvist, Britta Möller, Svante Forsberg, Mikael Rundqvist and the librarians for taking very good care of me.

Finally, I owe my greatest debt to my family, Marianne, Line and Esben. Although they paid the highest price for this book, they gave me much support. I also thank them for reminding me that life is much more than policy networks, pollution control and policy reforms.

Esbjerg, August 1997

Abbreviations

AE	Agra Europe
BEUC	European Bureau of Consumers' Associations
CAP	Common Agricultural Policy
COGECA	General Committee of Agricultural Cooperation in the European Community
COPA	Committee of Professional Agricultural Organisations
COREPER	Committee of Permanent Representatives
DDB	Danish Dairy Board
DFU	Danish Farmers' Union
DG VI	Directorate General for Agriculture
EC	European Community
EPA	Environmental Protection Agency
EU	European Union
GATT	General Agreement on Tariffs and Trade
LRF	Federation of Swedish Farmers

Introduction

Most Western states became deeply involved in agriculture in the early 1930s when the depression threatened many farmers' livelihoods. Food importing countries developed agricultural policies which restricted food imports in order to safeguard their farmers from the impact of the world market in which prices were rapidly decreasing. Food exporting countries responded to this protectionism by adopting agricultural policies which were to ensure the survival of their agricultural industries and to maintain farmers' income. Ever since these formative phases of agricultural policy making, Western states have continuously intervened in agriculture. In most West European countries, policy making has taken place in closed arenas in which farmers and state agricultural authorities controlled the policy field. The public and most members of parliament were excluded from participating in agricultural policy processes. Thus, agricultural policy networks existed almost in isolation from the rest of the political system.

Political developments in the 1980s and early 1990s forced agricultural policy makers to handle old problems in a new way and to deal with new policy issues. They welcomed neither of these developments. Problems which had been neglected or solved by channelling further economic resources into agriculture activated actors who did not traditionally participate in the agricultural policy process. Members of agricultural policy networks faced pressure for change because actors outside these networks became aware of the costs of subsidising agriculture and of the environmental damage farmers caused. Large surpluses caused budgetary problems, inefficient allocation of economic resources and trade problems in the world market. To cope with surplus production, states subsidised exports, either directly or indirectly. Thus, to a large extent, sales in the world market were the result of treasuries' ability to subsidise exports rather than of comparative advantages in agricultural production. Agricultural policies also had negative effects on the natural environment. Policies which subsidise farmers through high prices lead to intensive agricultural production which, in turn, causes environmental problems (e.g. OECD, 1989, pp. 29-30; 1993). The one receiving most attention in the 1980s and early 1990s was

1

nitrate pollution of ground and surface waters (Baldock, 1991, p. 13; Eckerberg et al., 1994, pp. 191-93). This is caused by the use of farmyard manure and chemical fertilisers.

Although Denmark and Sweden faced the same nitrate pollution problems, they chose different policy solutions. The Danes applied a low cost policy. Such a policy passes a large share of its political and economic disadvantages on to groups other than farmers. In contrast, the Swedes adopted a high cost nitrate policy which implies that farmers bear the economic and political costs of regulation. Similarly, Sweden and the European Community (EC)[1] faced similar types of problems in agricultural policies. Nevertheless, their agricultural policy reforms in the 1990s were fundamentally different. While the EC introduced more regulation into agriculture to cope with surplus production, Sweden deregulated the domestic market.

In abstract terms, this book deals with *policy change* in old, well-established policy sectors when political actors not traditionally participating in policy making in these sectors have succeeded in putting policy change on to the agenda. These agenda setters are outsiders. The concept of policy change, as used in this book, means reform of existing policies or the introduction of new policies designed to cope with problems which the established sectoral policy does not address. Policy reform refers to '... cases [in which] the change is [not] an entirely new departure for the State ... It constitutes a new way of doing something with which the State is already involved. Legislation or administrative structures are literally re-formed' (Hall et al., 1975, p. 19). The introduction of new policies 'calls for the entry of the State into a new field of social action or for the creation of new kinds of services, rights or obligations' (ibid.). Policy reforms and new policies can be designed in various ways and, thus, have different consequences for those subject to the changes. This means that policy changes have different images. The policy changes addressed here have a negative image to those who are subject to regulation because they detract from a situation which is conceived of as being advantageous. Farmers did not welcome agricultural policy reforms and the introduction of pollution control into agriculture because they believed such changes would disturb the established order in the agricultural sector and make them worse off.

How can we explain variation in policy changes? It is argued that the relationship between farmers and state actors which has developed since the early 1930s (since the late 1950s in the EC) is a major factor explaining variation. The relationship between organised interests and state actors can be characterised as a policy network. Networks emerge because actors within a policy field depend upon each others' resources. State actors need interest groups to provide information, to legitimise policies and to assist in policy implementation. Interest groups need state actors to have an influence on the contents of public policy. What is more, if relationships between state actors and organised interests develop into tight and closed policy networks, they then create stability and predictability in the policy process which is a major advantage for both state actors and interest groups. Policy network analysis deals with the relations between organisations and understands policy networks as structures (Marsh and Smith, 1996, pp. 12, 15).

Drawing attention to policy networks inevitably raises the questions: can policy network analysis help to explain why policy changes differ and what are the consequences of established networks when policy makers deal with policy change? This book argues that network analysis can help us understand why policy makers make different policy choices. *Network structures make a difference.* To give this statement force, we need to specify how established network structures influence the contents of both reforms and new policies in situations in which actors outside a network set the agenda. This research process involves a clarification of the concept of public policy in a policy network perspective.

The interest in explaining policy outcomes is not new within the network literature; however, although network analysts use the concept of policy outcome, they rarely specify what it means. To explain policy outcomes we need to define the concept. This book addresses the contents of public policies and attempts to classify them. This enables us to develop a theoretical model which establishes the causal link between network types and policy change. Most network studies are single case studies concerned with stability rather than change. By addressing the question of change in a comparative perspective, this book attempts to lead the way to comparative network analysis. At this point, it is important to address a crucial limitation of this book. The analysis deals with the way in which established policy network structures influence the contents of policy reforms and new policies when outsiders set the agenda. It does not analyse whether or not policy change leads to network change, neither does it analyse which types of policy networks can survive policy change. These questions are, however, important and need, therefore, to be examined in future research.

Some political scientists are less optimistic about the potentials of policy network analysis. While accepting that it is a useful tool for description, Dowding (1995, p. 145) criticises the policy network approach for being unable to provide explanations. He argues that:

> The nature of the policy process and the network of interests from which it emerges ... can be explained without recourse to the language of networks. ... This is not to deny that the policy network metaphor has no role to play, but it is to deny that it forms the centrepiece of explanation. Policies emerge through power struggles of different interests.

Dowding, thus, prefers an agency-centred approach. A similar analytical downgrading of political structures is also found in some contributions on policy communities. Although these distinguish between policy communities and issue networks (Jordan, 1990a; Jordan and Schubert, 1992), they do not attribute much explanatory force to network structures. As Jordan (1990a, p. 301) puts it: 'Political outcomes are the result of processes and not simply the consequence of structures.' Dowding is right that network analysis has not yet produced convincing explanations of policy outcomes, but he fails to see the potentials for developing a network theory. The point of departure for developing such a theory is the observation that different network types are associated with certain policy outcomes (Marsh and Rhodes 1992, 262; Rhodes and Marsh 1992a, 197; Stones, 1992, p. 224).

To understand why policy choices differ, we also need to analyse the broader context within which policy networks are embedded and within which sectoral policy processes take place (Marsh and Smith, 1996, pp. 12-13). Not until recently have network analysts begun to develop theoretical links between macro-variables and policy networks (see Marsh, 1995 for an example). Limiting this analysis to *macro-political variables*, I argue that social groups' structural power in parliament and the organisational structure of the state have a major influence on sectoral policy choices. The policy network literature does not show much interest in how parliamentary parties influence the formation and maintenance of sectoral (or sub-sectoral) policy networks, how they influence the stability of public policies over time or how they create opportunities for policy change in old sectors. Indeed, the policy network literature tends to view the political process as non-parliamentary (Brand, 1992; Judge, 1990, ch. 3; 1993, ch. 4). This is a serious mistake. Representative democracy is the major character of governance in Western societies. Therefore, it does not make sense to exclude parliamentary parties from the analysis. The other macro-political variable is less controversial. The organisational configuration of states encourages some types of collective actions and hinders others. Variations in the broader state structure, therefore, help to explain why public policies differ. Although macro-level analysis is important, the primary concern of this book is the impact of meso-level policy networks.

The main argument of this book is that fundamental policy change in an old, well-established policy sector is most likely to occur when the existing sectoral policy network is not cohesive, when the organisational structure of the state centralises authority and when the parliamentary support of the group subject to policy change is limited. In situations in which the sectoral network is cohesive, in which the state disperses authority and in which the group facing pressure for change has a high degree of support in parliament, policy change is likely to be modest or only marginal.

The origins of conflict between outsiders and farmers

More than six decades of state intervention in agriculture in Western Europe and North America did not, of course, occur without the development of strong regulatory traditions. State actors and farmer representatives developed policy principles which defined the contents of agricultural policy. The major principle concerned the state's role in agriculture. Western states have taken major responsibility for protecting farmers against economic imbalances. They have either guaranteed minimum prices for the most important agricultural products or given direct income support in order to guarantee farmers' incomes. Writing about the Common Agricultural Policy (CAP) of the EC in the early 1980s, Harris, Swinbank and Wilkinson (1983, p. 40) observed:

Market prices are administered, not competitively determined, and they are set at such levels that even high cost producers can earn a living. Within the CAP prices no longer have the limited role of bringing about an efficient allocation of resources and production; the implications for the incomes of farmers are judged to be unacceptable.

In other words, the EC has taken a significant responsibility for the economic well-being of farmers. What I shall call the *principle of state responsibility for economic imbalances in agriculture* has been, and still is, a major advantage to farmers. Without it, they would face unstable prices for their products and this would, in turn, cause unstable incomes and, perhaps, increase the number of bankruptcies. It would also make production management much more difficult because increased uncertainty in the market would require farmers to pay more attention to market forces when they plan their production. Although the Harris, Swinbank and Wilkinson quotation concerns EC agricultural policy, the same could be said about the agricultural policies of other Western states, in particular Sweden which stayed outside the Community until 1995. Swedish agricultural policy was, until 1991, based on the same model as the Common Agricultural Policy of the European Community.

In the 1980s, environmental problems caused by farmers came to the attention of environmental authorities and interest groups. These political actors were outsiders; that is, they did not belong to the agricultural policy network. They tried to introduce pollution control into agriculture. The result of this activity was that the established political order in the agricultural sector was seriously disturbed.

Since the late 1960s and early 1970s, pollution control has been on the political agenda in Western countries. Environmental authorities have also developed regulatory traditions. The basic principle in environmental policy making is the *polluter pays principle*. It was first adopted by the OECD member states in 1972 and by the EC in 1973 (OECD, 1989a, p. 27; Baldock, 1991a, p. 15). In 1986, the Single European Act amended the Treaty of Rome so that the polluter pays principle became part of European law (Baldock, 1991a, p. 17). However, countries have interpreted the principle in various ways, allowing for different types and levels of state subsidies (OECD, 1989a, pp. 27-30). Denmark and Sweden have officially recognised that the principle should be applied in environmental policy making (Lov no. 372, 1973; Andersen, 1994, p. 83; Christiansen, 1996, p. 32; Proposition 1987/88: 85, p. 42).

Basically, the polluter pays principle is incompatible with the dominant principle in agricultural policy making, the principle of state responsibility. Applying the latter principle invalidates the former, because then the state, and not the polluters, would pay the costs of pollution control. Therefore, when pollution control was put on to the aganda, a political conflict emerged over what principles to apply. Agricultural policy had a positive image for farmers because the principles on which it was based benefited them politically and economically. Perhaps more importantly, applying the principle of state responsibility had become an accepted way of solving agricultural problems. Hence, farmers viewed the efforts of environmental interests to introduce a new policy issue into the agricultural sector in a negative light (see Baumgartner and Jones, 1993, pp. 81-89) and opposed agri-environmental policies based on the polluter pays principle.

Serious pressure for reforms of agricultural policies also occurred in the late 1980s and early 1990s. In both the EC and Sweden, political actors who did not traditionally participate in agricultural policy making tried to force more *market principles* into agricultural policies. The pressure upon the EC to reform its Common Agricultural Policy came from the GATT Uruguay round and from yet another imminent budgetary crisis. In the Uruguay round which started in 1986, the United States and the Cairns Group, a group of 14 'fair trading' nations,[2] put heavy pressure on the European Community to liberalise its agricultural policy. The Americans threatened a trade war (*AE* no. 1421, 4 January 1991, E/1). Furthermore, by the end of 1990, the recurrent problem of the CAP - a forthcoming budgetary crisis - once more loomed. If nothing had been done, the crisis would have been the most severe ever in EC history. In Sweden, there was also considerable pressure for more market orientation in agriculture. The minister of finance, Kjell-Olof Feldt, and the Department of Finance pointed out that the agricultural policy caused welfare losses and inflation (Feldt, 1991, p. 343; Finansdepartementet, 1988, pp. 35-38). When they initiated the 1990 agricultural policy reform, these actors wanted deregulation as a means to cope with the problems of agricultural policy (Feldt, 1991, p. 346).

The Swedish and the EC reform processes of 1990 and 1992 respectively involved a clash between the principles of state responsibility and of the market. Consequently, a political conflict emerged. Farmers defended the existing agricultural policy because they conceived of it as positive. Because deregulation and the introduction of more market forces into the agricultural sector created uncertainty and income instability, farmers opposed agricultural policies based on the market principle.

An obvious question to ask is: can differences in the external pressure on agricultural policy makers in agri-environmental policy making and agricultural policy reform explain why policy choices differed? Clearly, the answer is *no*! The annual nitrate run-offs have been estimated at 250,000 tonnes in Denmark and 50,000 tonnes in Sweden (Eckerberg et al., 1994, p. 191). Had the seriousness of the nitrate pollution had an impact on the choice of policy, then the Danish nitrate policy should have been the more radical. Since this was not the case, there must be other explanations (Daugbjerg, 1996). It is difficult to assess whether there were different degrees of external pressure in the two cases of agricultural reform. Nevertheless, one could argue that the American threat to begin a trade war with the EC showed that the external pressure on the Common Agricultural Policy was greater than that on Swedish agricultural policy. Thus, if differences in external pressure explain reform outcomes, EC policy reform should have been the more fundamental, and it was not.

Did policy choices differ?

This book examines which policies governments chose to cope with the problems raised by outsiders. Were nitrate policies based upon the polluter pays principle or was the principle of state responsibility just transferred from established agricultural policy to nitrate policy? Did policy makers respond to the underlying economic problems of agricultural policy by fundamental reforms, or were the changes which actually occurred only adjustments of the existing policy?

The Swedish nitrate policy is a high cost policy whereas the Danish nitrate policy is a low cost policy. Swedish farmers bear a larger share of the political and economic costs of environmental regulation than do their Danish counterparts. For instance, since 1984, the Swedes have applied fertiliser taxes. Danish environmental interests advocated similar measures, but Danish farmers successfully resisted several attempts to introduce them. The choice of policy objectives also shows that politically, Danish farmers were the more successful. The objectives of Swedish nitrate policy are more constraining than those of the Danish. In Sweden, the policy sets a specific reduction goal for the consumption of fertilisers, whereas such a goal does not exist in Denmark. These differences indicate that Swedish farmers were less successful than their Danish counterparts in transferring the main principle of agricultural policy making (the principle of state responsibility) to nitrate policy making. Hence, Danish nitrate policy is, to a large extent, based on the principle of state responsibility whereas the Swedes, by and large, base theirs on the polluter pays principle.

The empirical examination of agricultural policy reforms in Sweden and the European Union shows that the 1990 Swedish reform was a fundamental one while the 1992 EC reform was moderate. Swedish policy makers decided to rely more on market forces in agriculture and policy makers in the EC responded to the problems with new regulatory measures. To limit surplus production and reach agreement with the Americans in the Uruguay round, the EC switched from a high price to a low price policy combined with set aside schemes in the arable sector. Now, the EC pays most subsidies in the arable sector directly to farmers through acreage payments. To qualify for acreage support, the farmers must set aside land. Before the reform, farmers were subsidised through high consumer prices. To sum up, the reforms of the Common Agricultural Policy have 'introduced new instruments for the management of policy, but they have not fundamentally changed the nature of the policy itself' (Grant, 1995, p. 2; see also Manegold, 1993, p. 126). Thus, the CAP is still based on the principle of state (or EC) responsibility for economic balances in agriculture.

The Swedes deregulated their agricultural policy in 1990. While keeping up border protection, they chose to abolish the domestic market regulation. Moreover, new policy objectives removed the justification for subsidised surplus production. Thus, Swedish policy makers, to a large extent, based the 1990 agricultural policy on market principles rather than on that of state responsibility. However, the reform was not fully implemented. Swedish application for EC membership brought the implementation process to a halt. She joined the European Union in 1995 and is now implementing the Common Agricultural Policy and, therefore, the national policy reform will not be carried out. However, this does not mean that political scientists should lose interest in the reform. From an international perspective, it is remarkable that Swedish policy makers enacted a fundamental change. Compared to the EC and other Western countries, the Swedes chose a radical solution. Only New Zealand has applied similar measures. She deregulated the agricultural policy in 1984 (see e.g. Le Heron, 1993; Cloke and Le Heron, 1994; Grant, 1991). These two countries are, therefore, valuable cases in international comparisons of agricultural reform processes since they represent contrasts to the responses of most other countries. These, and the EC, have reacted to the problems of agricultural policy by implementing further regulations. For a political scientist, the Swedish case may even be more interesting than that of New Zealand. While, in the final analysis, we can ascribe the policy change in New Zealand to the country's economic crisis in the 1980s (Grant, 1991, pp. 63-65), political and organisational factors are much more important in understanding the Swedish reform.

Why did policy makers choose different policies?

We can only understand variation in policy choices within a theoretical perspective. Reality is extremely complex. It is not possible to explain events without some consideration of causal relations before we undertake analyses of real-life phenomena. It is, therefore, necessary to develop theoretical models. A theoretical model is 'an abstraction of reality that is developed for presenting systematically the most important relationships in the situation which is being described' (Graham, 1971, p. 112). It is 'highly suggestive to the researcher because it points the way to significant research' (ibid.). Since a model is only suggestive, it cannot, of course, be a generally valid theory on a phenomenon before it has been systematically tested empirically. A model is an analytical tool which is generally applicable to the phenomenon it addresses. Models should, therefore, use general terms and concepts.

In brief, the theoretical network model (chapter 2) suggests that the existence of a cohesive policy network in the sector in which reform or the introduction of a new policy is put on to the agenda limits the opportunities for fundamental policy change. In such a network, members have developed a consensus on the principles to underpin the choice of policy and on the way to handle policy problems. The consensus means that the network's core members can form a strong status quo minded coalition when outsiders try to bring about policy change. The existence of strong opposition leads the policy process in the direction of moderate policy reforms and new policies tend to be low cost policies. Where a non-cohesive network exists, there are better opportunities for fundamental policy change. Lack of a consensus on policy principles and on the ways of handling new problems in such a network impedes the formation of a strong coalition which can oppose change. In these circumstances, outsiders have favourable opportunities for bringing about radical policy reforms and new high cost policies. So, the character of networks is a crucial variable explaining the contents of policy changes.

The analysis of agricultural policy networks demonstrates that the model has predictive force in agricultural policy reforms and in agri-environmental policy making. Swedish farmers could not mobilise sufficient support to prevent the adoption of a fundamental agricultural policy reform and a high cost nitrate policy because the agricultural network has a low degree of cohesion. The Swedish agricultural policy network is unique. While the cores of agricultural policy networks, in general, consist of two parties (farmer associations and ministries of agriculture), the Swedish network is a tripartite body. It consists of farmers, agricultural state actors and consumers. Consumers are organised in the Consumer Delegation which became a member of the network in the early 1960s. This unique construction is an important reason for the absence of cohesion and explains why Swedish agricultural state actors did not form a coalition with the status quo minded farmers. Methodologically, the Swedish case is valuable. It represents an extreme example of Western European agricultural policy networks. Sweden is the only case in which farmers' counterpart in the market (the consumers) is strongly represented in the agricultural policy process. Usually, consumer influence is marginal. A comparison of Swedish agricultural politics with agricultural politics in countries with two-party agricultural policy networks suggests that we can test theoretical models stating the conditions under which a particular outcome occurs or not. The logic underlying this type of multiple case study design is called theoretical replication. It provides the most favourable conditions for the generalisation of research results (Yin, 1989, pp. 52-54, 109-112).

In Denmark, by contrast, the Ministry of Agriculture became an important ally of farmers. The core of the EC agricultural policy network is not quite as cohesive as the Danish network but, in contrast to Sweden, it does not include a countervailing power. In the EC agricultural policy reform process, the Directorate General for Agriculture (DG VI) and the agriculture commissioner safeguarded the basic interests of farmers. Danish and EC farmers could limit the effects of the external pressures for policy change because the existence of cohesive networks enabled them to protect their interests.

Organisational arrangements at the meso-level cannot fully account for variation in policy choices. Chapter 3 points out that macro-political variables also affect the contents of both policy reforms and new policies. A social group's support in parliament influences outsiders' opportunities. If the interest groups representing those subject to policy change can mobilise support in parliament, then policy reforms are likely to become moderate and new policies will be low cost. Where the parliamentary support of a social group has eroded considerably over time, outsiders have favourable opportunities to introduce new high cost policies and bring about fundamental policy reforms. Further, it is argued that the organisational structure of a state (or a political system) influences the contents of meso-level policy changes. In states with centralised political decision making structures, reformers have favourable opportunities for successfully pursuing radical policy change because political power can be centralised in one body. Once such a body decides to pursue fundamental policy shifts in a sector, it is, to a large extent, able to ignore political opposition. In other words, there are few, if any, veto opportunities which status quo minded actors can use to prevent change. States (or political systems) in which the structure disperse political power to several decision making centres provide many veto opportunities which can often successfully be used to mobilise opposition to change.

The empirical analysis shows that agricultural policy reformers and environmental interests in Sweden could derive political power from macro-political conditions, whereas macro-political factors in Denmark and in the EC favoured agricultural interests. In the Swedish parliament, the power of farmers has declined since the formative phases of agricultural policy in the 1930s. Danish agricultural interests still enjoy considerable support in parliament. This difference helps to explain why Danish policy makers chose a low cost nitrate policy and the Swedes adopted a high cost policy and also reformed fundamentally the agricultural policy. The comparison of state structures shows that Denmark and Sweden have a high degree of centralisation. Thus, state structures did not help to explain why the two countries' nitrate policies differ. Agricultural reform attempts were seriously constrained by the highly fragmented structure of the EC. In reality, the EC (until the entry of Austria, Finland and Sweden in 1995) had 13 decision making centres: the EC Commission in Brussels and twelve national governments. The latter are strong decision making centres. In practice, they each have power to supplement the Common Agricultural Policy with national support measures if they are not satisfied with EC measures. Indeed, all EC countries have national agricultural policies which supplement the Common Agricultural Policy. Since the influence of the European Parliament in agricultural policy making is marginal, the macro-level analysis does not focus on farmers' parliamentary support at the EC level. To sum up, the pursuit of radical reform was much more difficult in the EC than in Sweden because the EC's broader organisational structure allowed many veto points to exist.

As argued above, variation in policy choices is mainly explained by political and organisational variables. Inevitably, this raises questions about the role of macro-economic factors. Are they part of the explanation? Macro-economic factors may help to explain why an issue appears on the political agenda, but they cannot explain why governments choose certain solutions and not others. Even in New Zealand, where a major balance of payments and fiscal crisis triggered the agricultural policy reform process, political and organisational conditions had a vital influence on the choice of deregulation as the best policy solution (Grant, 1991, p. 64; Roche et al., 1992).

Economists and some political scientist have often suggested that the importance of agricultural production in the national economy explains why governments choose different policies. For instance, Bennet and Baldock (1991, pp. 221-2) use mainly non-political variables to explain differences in agri-environmental policies. They suggest that the physical environment, the agricultural structure and intensity, the economic importance of agriculture and, finally, the political strength of farmers explain why policies differ. All four variables do, of course, play some role. Bennet and Baldock do not rank them according to their explanatory power. If they did so, they would realise that political factors are the most important. There need not be a correlation between economic importance and the costs which state intervention inflict upon farmers. The economic proposition suggests that in countries in which agricultural production is important for the national economy, policy makers are careful not to decrease the international competitiveness of agriculture (see e.g. ibid., p. 222). Consequently, they choose solutions which do not burden farmers economically. However, the national economic importance of agriculture cannot explain policy choices. Two comparisons support this statement.

Firstly, agricultural production in Norway and Sweden is almost equally important for the national economies of the two countries. Agriculture accounts for 1.8 per cent of the GDP in Norway (1990) (Statistisk Sentralbyrå, 1993, p. 335) and 1.5 per cent in Sweden (1990) (Swedish Board of Agriculture, 1994, p. 3). According to the economic proposition, we should expect the two countries to choose relatively similar policy measures, but they have not done so. The level of economic agri--environmental subsidies in Norway is much higher than in Sweden (Nordisk Ministerråd, 1992, vedlegg 1, pp. 2-3; Eckerberg et al., 1993, pp. 198-9). Secondly, although the agricultural sectors in New Zealand and Sweden play different roles in the national economies, the two countries chose fairly similar solutions to overcome the problems of agricultural policy. While agricultural production accounts for only 1.5 per cent of GDP in Sweden, it makes up 6.9 per cent (1990) in New Zealand (Sandrey and Reynolds, 1990, p. 2). The economic explanation predicts significantly different outcomes. So, empirical evidence demonstrates that the economic explanation is, indeed, not a valid explanation. To qualify as a theoretical explanation, similar conditions must produce similar outcomes (which they do not in the comparison between Norway and Sweden) and different conditions must produce different outcomes (which they do not in the comparison between New Zealand and Sweden). The reason why the economic proposition has been so influential in explaining policy choices in agriculture may be that it has not been subject to comparative tests in which the researcher chooses cases strategically.

These results do not, however, prevent political scientists from applying economic variables. Since economic importance is not objectively given, we should examine how political actors perceive the economic role of an industry and how their perceptions influence policy processes. As Smith (1990a, p. 37) points out: 'A pressure group or interest can increase its [political] influence by convincing the government that it is economically important.' How policy makers perceive the economic importance of an industry, therefore, has important consequences in politics.

The findings of this book go beyond the interests of those studying policy networks and political institutions. The analysis also addresses matters which concern policy analysts and policy designers. By paying attention to the process by which policy measures are chosen, this book explains part of the frustration of many policy analysts and implementation researchers. They seem disappointed by the lack of political support for their policy recommendations although they claim that such recommendations are optimal in terms of efficiency and effectiveness.

The comparative case study method

Research results should always be assessed in the light of the applied research methods. Therefore, this section describes how data was collected and analysed. Further, the question of validity in comparative case studies is addressed. The case study method can be used for various purposes within the field of political science. For instance, we can apply it in exploratory, descriptive and explanatory studies. A case study is 'an empirical inquiry that investigates a contemporary phenomenon within its real-life context; when the boundaries between phenomenon and context are not clearly evident; and in which multiple sources of evidence are used' (Yin, 1989, p. 23). The method requires that the researcher see political life as a whole, stressing the uniqueness of the cases analysed but, at the same time, focusing on generic concepts (see Ragin, 1987, ch. 3; Dogan and Pellasy, 1984, pp. 112-4; Rose, 1991, p. 453). Usually, we use the case study method to analyse cases in which we are dealing with data which cannot be quantified, for example, statements appearing in documents and information obtained from interviews. To understand nitrate policy making and agricultural policy reforms, we must analyse non-quantifiable data. This data is, primarily, collected from documents and interviews. Therefore, the only suitable research method is the use of case studies.

The comparative case study method is based on the logic of multiple experiments. When researchers use such methods, they obtain insights into a phenomenon by replicating a series of experiments. The more times the experiment produces the results predicted by theory, the more convincing the theory becomes. Multiple experiments can be designed as literal replications or as theoretical replications. In experiments based on literal replication, the researcher designs experiments which are predicted by theory to produce similar outcomes. In theoretical replications, s/he designs experiments which for predictable reasons produce different outcomes. Applying the same logic in case studies, we can select cases which the theory predicts will produce either similar outcomes or different outcomes. We can also design case studies which employ both types of logic (Yin, 1989, pp. 53, 109-10). Comparative case studies using both literal and theoretical replications give strong support for theoretical explanations. Both designs are applied in the four cases.

The meso-level theoretical model suggests that cohesive policy networks produce policy responses to external pressure which are different from those of non-cohesive policy networks. I base the comparison of nitrate policy making in Denmark and Sweden on the logic of theoretical replication because the theoretical model predicts that the two policy processes produce different outcomes. My proposition is that Denmark has a cohesive agricultural policy network, whereas the Swedish network has a low degree of cohesion. Similarly, the theoretical model predicts that agricultural policy reforms in the European Community and Sweden will be different because the EC agricultural policy network seems to be more cohesive than the Swedish. Literal replication is also involved in the comparison since I analyse two cases in which political actors not traditionally participating in agricultural policy making confront cohesive policy networks (nitrate policy making in

Denmark and agricultural policy reform in the EC) and two cases where they face networks with a low degree of cohesion (nitrate policy making and agricultural policy reform in Sweden). With regard to Denmark, the theoretical model predicts that the established agricultural policy principles will underpin Danish nitrate policy. In the European Community established agricultural policy principles will remain the basis of the Common Agricultural Policy. In contrast, the theoretical model predicts that new policy principles will underpin the Swedish nitrate policy and the reformed agricultural policy. The analysis of macro-political factors is based upon a similar design.

Analysing data and the question of validity and reliability

Case studies must have a high degree of validity and reliability in order to contribute to the development of theory. Validity refers to whether there is a link between the theoretical question and the empirical analysis. Reliability refers to whether repeated independent 'measurements' by the same method will yield the same approximate results (Hellevik, 1987, p. 155; Kvale, 1989, p. 79; Yin, 1989, p. 45). Yin (1989) subdivides the concept of validity into construct, internal and external validity. The latter is addressed in the final section of this chapter.

Construct validity is obtained when correct operational measures are established for the theoretical concepts applied (Hellevik, 1987, p. 157, Kvale, 1989, p. 75, Yin, 1989, pp. 40-42). Usually, we deal with latent phenomena which do not reveal themselves to the researcher, for example the concept of power. Researchers need to develop a manifest expression for the phenomenon to measure it empirically (Hellevik, 1987, p. 157). A simple manifest expression of power might be: who makes the decisions? We can obtain construct validity by using multiple sources of evidence (Yin, 1989, p. 97). This technique is called triangulation (Kalleberg, 1982, p. 32, see also Murphy, 1980, p. 71). In this book, written sources of evidence were analysed and interviews were conducted. Findings from one source must agree with the findings of the others. The results obtained have, if possible, been cross-checked with other authors' findings and conclusions in order to assess the construct validity. Another way of obtaining construct validity is to have key informants review drafts of case study reports (Yin, 1989, pp. 143-5, Buchanan et. al., 1988). However, although drafts were sent to most of the people interviewed in 1993 and 1994, I received comments from only two of them. This method was not, therefore, successful. Using several operational definitions of the same theoretical definition is a third way to increase construct validity. A high degree of agreement between the empirical findings in different operational variables increases the construct validity (Hellevik, 1987, pp. 168-9). That technique has been employed in this book.

Internal validity is obtained when a causal relationship is established; that is, when we can show that certain conditions lead to certain outcomes and alternative explanations can be ruled out (Yin, 1989, p. 40, see also Hellevik, 1987, pp. 346-63). In the beginning of the research process, it was not possible to stipulate a set of theoretical propositions. In such a situation, Yin (1989, pp. 113-5) suggests that we use the method of explanation building which builds theoretical explanations step-by-step. To obtain internal validity, this method uses a series of iterations, involving several stages:

- making an initial theoretical statement or an initial proposition about the policy or social behaviour;
- comparing the findings of *an initial case* against such a statement or proposition revising the statement or proposition;
- comparing other details of the case against the revision again revising the statement or proposition;
- comparing the revision to the facts of *a second, third or more cases*;
- repeating this process as many time as needed (ibid., pp. 114-5).

The process of explanation building ends when the explanation is free of inner contradictions (see Kvale, 1983, p. 185). During the research process, theoretical explanations were matched against plausible and rival explanations. When it was possible to identify reasonable threats to internal validity, further analysis took place to revise the theoretical explanation or to reject alternative explanations. Unfortunately, there is no precise technique which can be applied to assess the internal validity of rival explanations. Until we develop precise techniques, 'an "eye-balling" technique is sufficiently convincing to draw a conclusion' (Yin, 1989, p. 113). Scholar triangulation, i.e. having other researchers review the interpretations (Kalleberg, 1982, p. 32), can also increase internal validity. Several scholars were involved in that process (see the acknowledgements).

Reliability is obtained when other researchers can reach the same results by the use of the same methods. Basically, reliability is about avoiding biases and errors. The best way to ensure reliability is to maintain a chain of evidence; that is, to account for procedures and to make sufficient references to books, articles, documents, interviews, etc. which have been used as sources of evidence. In principle, the reader of a case study should be able to follow the research process from the initial questions to the final conclusions (Yin, 1989, pp. 45, 102). This book attempts to maintain the chain of evidence by accounting for the research design, by referring extensively and precisely to sources of evidence and by explaining research procedures.

In many cases, researchers have to trade off validity for reliability, or vice versa. An accurate measure for a theoretical phenomenon may be reliable, but may have a low degree of validity. On the other hand, a valid operational definition may not be sufficiently reliable. In political science, we have to live with that problem because the most interesting theoretical variables are often those for which it is most difficult to develop operational definitions (Froman, 1968, p. 45).

Analysing the sources of evidence

Validity and reliability are obtained through the analysis of the sources of evidence. Therefore, the methods applied must be explicit to critical readers, so that they can verify the research results (Kvale, 1984, p. 57). Several sources of evidence have been applied in this study, particularly public documents and second hand literature. Interview data was used to some extent. Archival records and direct observation have only been used to a limited extent. Therefore, I limit the account of methodological issues to documents, second hand literature and interviews. Ideally, a case study should examine all the relevant sources to which the researcher can gain access; however, in practice, s/he is forced to choose only the most relevant ones due to resource constraints (Dahl, 1980, p. 50).

Public documents are a major source of evidence in the analysis of agricultural policy reforms and of nitrate policy making. Such documents can be used for two purposes. Firstly, documents tell us something about the situation in which they were written. They give information about the author's motives, perceptions, feelings, values, assessments, etc. A document used in this way may also tell us something about the author's reasons for writing it and why s/he wrote it in a particular way. To apply documents as sources of evidence in this way, we need to take the role of the author into consideration. Does s/he write as a member of an organisation or as a private person? Is the source confidential or public? Secondly, some documents tell us about a historical event. There are certain research procedures to consider when we use documents to give us information about the past. Essentially, we want to assess whether a source of evidence is relevant and trustworthy. In analysing first hand and second hand sources, we should generally choose the one which comes closest to the event in time and the one in which the author was physically closest to the event. In assessing a source of evidence, it is important to examine the provenance of its information. If it is based on other sources, we should try to gain access to these. What is more, the information from one source should agree with that of other sources (Dahl, 1980). In dealing with first hand sources of evidence, the purpose is to reconstruct a historical reality and to develop theories which can explain why the sources are as they are (ibid., pp. 34, 51).

Second hand literature plays an important role in the analyses of policy networks and macro-political factors. Whether books and articles are relevant and trustworthy sources of information depends upon their reliability, construct validity and internal validity. The above section deals with techniques to assess these indicators of relevance and trustworthiness. All books and articles used in the empirical analyses have been checked for reliability, construct validity and internal validity.

Interviews are another important source of evidence in this book. I have conducted qualitative research interviews. This type of interview has been characterised as 'semi-structured' (Kvale, 1983, p. 174), 'intensive' (Murphy, 1980, pp. 75-77), and 'informal' (Hellevik, 1987). In an informal (or semi-structured or intensive) interview, questions are presented unsystematically. Registration of data is also unsystematic; that is, the questions have no fixed categories of answers. The advantage of the informal interview is that it leaves the investigation open to new and unexpected information (Hellevik, 1987, pp. 100-07). Interviewees can be informants or respondents. In the former role, they are witnesses who can give information on, for instance, what happened and in what sequence, institutional norms, perceptions and motives. If we interview a person as a respondent, s/he becomes a representative of the social group, firm or organisation to which s/he belongs. In such a case, we are interested in personal opinions, perceptions and motives (Kalleberg, 1982, pp. 24-25). In the interviews which I conducted, interviewees were both informants and respondents. The interviews focused on how policy makers made decisions and how the interviewees, as representatives of organisations, conceived other actors, what motivated their actions and which interests they pursued. In order to obtain construct validity, I cross-checked interview data with data from other interviews and other types of evidence.

A major purpose of the interviews was to reveal the perceptions of persons representing various organisations. The interviewer 'attempts to condense and formulate what the interviewee himself understands as the meaning of what he describes' (Kvale, 1983, p. 181). Clarifying the meaning of an interviewee's answers is often made easier by using leading questions (Kvale, 1984, p. 58). The purpose of these is not to lead the interviewees to express certain opinions, but to lead them in the direction of certain themes (Kvale, 1983, p. 190). Leading questions have been used as an 'on-line-interpretation' of the interview. Interviewees can confirm or reject one's initial interpretations and, consequently, make the subsequent analysis more precise.

Case studies and the question of generalisation

Can the case study method produce findings which reach beyond the specific case which is studied? If it can, how do we then generalise? Yin (1989, p. 43) refers to the problem of generalising as the problem of external validity. In surveys, generalisation is based on statistical methods by which samples are selected so that they mirror the populations or universes from which they are drawn. Yin (1989, p. 43) argues that the 'analogy to samples and universes is incorrect when dealing with case studies.' Instead, Yin (1989, pp. 21, 44) stresses that the case study method relies on *analytical generalisation*. Case studies are generalisable to theoretical propositions. The researcher generalises the results in relation to the theory which addresses the phenomenon in focus. In case studies, it is from the development of theory that we can obtain knowledge on social and political phenomena and not from statistical methods. Development of theory is, therefore, the main tool of generalisation in case study research.

Generalisations will become more compelling the more cases we investigate. Multiple case studies based upon the logic of literal replication can strengthen theoretical propositions; however, they do not support theories which attempt to establish the way in which variation in the independent variable is related to variation in the dependent variable. Therefore, multiple case studies based upon theoretical replications provide better opportunities for theoretical development. They test whether empirical evidence supports the variation predicted by theory. Theoretical replications thus produce results which, in terms of generalisation, are more convincing than those of literal replications.

Outline of the book

Chapter 1 reviews four network approaches and discusses whether the policy network concept should be conceived of as a metaphor for a pattern of interaction or whether it makes more sense to view the network itself as a crucial determinant of policy. The chapter concludes that applying the network concept as more than a metaphor is analytically meaningful. The review reveals a general problem in the policy network literature. None of the four approaches has yet been able to develop a theoretical model which can explain the choice of policy. Chapter 2 develops a network continuum which consists of dimensions describing the characteristics of a network and it develops a model linking networks with policy change. Chapter 3 develops the macro-level model (both models are briefly described above). Chapter 4 conceptualises and classifies policy reform and environmental policy. It also sets out the propositions on which the empirical analysis is undertaken. Chapter 5 compares nitrate policy making in Denmark and Sweden, and chapter 6 is a comparison of agricultural policy reforms in the EC and Sweden. Chapter 7 compares agricultural policy networks in Denmark, Sweden and the EC. In chapter 8, I undertake a macro-level analysis which compares Danish and Swedish farmers' structural power in parliament. I also compare state structures in Denmark and Sweden with the organisational structure of the EC. Finally, chapter 9 draws conclusions from the study.

Notes

1. There is some confusion about the terms European Community and European Union. Policy making within the first of the three pillars of the European Union is within the European Community. The first pillar deals with most policy areas, e.g. agricultural policy, the single market project and competition policy. The second pillar concerns foreign policy and the third one legal affairs. Since agricultural policy making is within the European Community, I use that term here.

2. It consists of Argentina, Australia, Brazil, Canada, Chile, Columbia, Hungary, Indonesia, Malaysia, New Zealand, the Philippines, Singapore, Thailand and Uruguay.

1 Policy networks: a critical review

Introduction: research traditions within the literature on government-interests relations

The relationship between government and organised interests has preoccupied political scientists for almost a century. Arthur Bentley, who in 1908 published *The Process of Government*, is often regarded as the founder of interest group studies. He demonstrated that such groups played an important role in the governmental process. Subsequently, hundreds of studies of government-interest group relations in industrialised democracies have been undertaken. We can identify four broad traditions within this research: pluralism, sub-government studies, corporatism and policy network analysis. Bentley was the first to write in the pluralist tradition and many followed him. In the 1950s and 1960s, pluralism was challenged in the United States by the literature on iron triangles, sub-systems and sub-governments. In Europe, pluralism came under heavy attack from corporatism in the 1970s; however, in the mid-1980s corporatism went out of fashion. Policy network analysis is the most recent tradition within the government-interest relations literature. Many view it as an attempt to overcome the shortcomings of corporatism and classical pluralism.

The basic argument of classical pluralism is that, in the long run, interest representation exhibits some sort of equilibrium. Many pluralists view access to the policy process as being fairly open. Despite the notion of balanced interest representation and of openness, classical pluralists accept that political power is distributed unequally. Political power is not, however, concentrated in the hands of the few; it is dispersed among a large number of groups. Overlapping membership also constrains any interest group's exercise of power. If a group becomes too powerful, it may provoke internal opposition because some of its members may also be members of other groups which suffer from this use of power. These may leave the group and form a new, opposing association. Excessive power may also bring about opposition from unorganised groups which also perceive the activities of a powerful group as being too costly. The result is that such groups will mobilise politically. These factors make powerful interest groups moderate their political demands (Truman, 1951, ch. 2, 3, 6). In addition, an interest group's use of power is constrained by government officials who, in an attempt to balance interests, act on behalf of unorganised and potential groups (ibid., pp. 448-9). In classical pluralism, the main function of an interest group is to aggregate the interests of its members and represent them to the policy makers (Almond and Powell, 1966, ch. 4).[1]

The picture presented by classical pluralists was challenged by plural elitists who argued that the policy process is much more closed than was assumed in classical pluralism. They '[admit] the absence of a single power elite but [view] public policy as fragmented into hundreds of separate political arenas, many of them under the control of particular elites' (McFarland, 1992, pp. 60-61). Plural elitists claim that the countervailing power of potential and unorganised groups is limited; some groups remain more powerful than others (ibid., p. 58).[2]

The European response to classical pluralism was different. Schmitter's article on corporatism, published in 1974 (Schmitter, 1974), set a new agenda for studies of government-interest relations by arguing that corporatism characterised interest representation in most European nations.[3] In corporatism, interest groups change from being primarily representatives of special interest into being interest intermediators. As such, their role is to mediate between the state and their group membership and not to represent the interests of their members only. In Williamson's (1989) words:

> A major function of interest association under corporatism is to regulate their members' articulation of interests and to ensure a measure of compliance with public policy, and that this is done to some degree on the behalf of the state's interests, as opposed to the collective interests of the members (1989, p. 117).

To perform this intermediatory function, an interest association requires a certain control over its members so that agreements with the state can be implemented. Pluralists accept that interest groups can carry out regulatory functions on behalf of the state as long as they regulate by consent. In contrast, corporatists argue that regulation is often carried out by imposition (Williamson, 1989, pp. 103-4). Interest associations become intermediators in corporatism because they exercise a certain control over their members and because the state grants them representational monopoly.

Corporatism has been strongly criticised. Firstly, there was significant doubt whether or not it was distinct from advanced forms of pluralism (Jordan, 1981; 1984; Almond, 1983, p. 251), especially from what has been labelled corporate pluralism which also observed the regularised incorporation of organised interests into the public policy process (e.g. Heisler and Kvavik, 1974, pp. 43-63). Secondly, corporatism suffered from conceptual confusion. It is not clear what its object was or what it attempted to explain. Corporatism was for example seen as a system of interest intermediation; a form of policy making; an economic system; a device for conflict management in advanced capitalist societies (Wilson, 1983, pp. 106-10) or a state form (see Smith, 1993, pp. 29-31). Thirdly, it has not been possible to underpin corporatism empirically (Williamson, 1989, pp. 197-200; Wilson, 1983, p. 117). For instance, the order predicted by corporatist theory was not found in extensive empirical studies of the Nordic countries (e.g. Heisler, 1979, pp. 284-6; Buksti and Johansen, 1979; Buksti, 1983a) which were believed to be vulnerable to corporatism. Reality was far more complex than corporatist theory claims. Although corporatism had serious shortcomings, it has provided important contributions to research on government-interest group relations. Firstly, it encouraged 'those studying interest groups to go beyond the usual concern with the internal organization of the group to explore its relations with the state and its impact on policymaking' (Wilson, 1983, p. 120). Secondly, by focusing on the exclusion and inclusion of groups in public policy making, it questioned pluralists' notion of some sort of long term equilibrium in power relations, reminded us that power could be unequally distributed and that the conditions for the formation of countervailing forces need not always be present. Thirdly, it challenged positivist research methods and methodological individualism. Finally, and perhaps most importantly, corporatism drew attention to the role of interest group leaders in public policy making. Unlike pluralists, corporatists argued that such leaders had devices for disciplining rank-and-file members.

In a response to the massive criticisms of corporatism, Cawson (1985) tried to save the concept by developing a meso-corporatist approach. He argued that although corporatist arrangements might be weak or non-existent at the national level, they might appear at the sectoral level (Cawson, 1985, p. 2). Cawson's attempt to develop the concept of meso-corporatism was, however, not enough to prevent corporatism going out of fashion in the late 1980s.[4]

The policy network tradition evolved partly out of the debate between corporatism and pluralism. In particular, the inability of both to account for variation between policy sectors proved an important starting point for, and focus of, network analysis. In order to provide a more realistic picture of government-interest relations, network analysts disaggregate policy analysis to policy sectors or areas; stress variation between sectors; recognise that in many policy areas only a limited number of interests are involved and emphasise that there is continuity in membership of policy networks in many sectors (Rhodes and Marsh, 1992, pp. 3-4). Although we can speak of a network tradition, it is not coherent and has taken different directions. Here, I deal mostly with the British discussion on policy networks which concerns fundamental questions within political science. The debate, therefore, reaches beyond policy network analysis.

In this chapter, I review four different network approaches in the light of the emphasis each of them puts on structure and agency respectively in explaining policy outcomes. The main question is: can policy network structures explain policy outcomes? The chapter attempts to establish whether we should perceive a policy network as a metaphor for a pattern of interaction or give it explanatory force by emphasising that the network structure is an important determinant of policy outcomes. The critical review concludes that the policy network structure itself is an important variable influencing policy outcomes.

The policy network literature: what holds it together and what divides it?

The policy network tradition tries to overcome the rigidities of both classical pluralism and corporatism. It is held together by a common definition of the policy network concept. Following Benson (1982, p. 148), many network analysts define a network as: '... a cluster or complex of organizations connected to each other by resource dependencies and distinguished from other clusters or complexes by breaks in the structure of resource dependencies' (quoted in e.g. Rhodes, 1990, p. 304, Jordan, 1990a, p. 319; Wilks and Wright, 1987, p. 299; Wright, 1988, p. 606; Smith, 1993, p. 58; Rhodes and Marsh, 1992, p. 13). Resource interdependency is the most important defining feature of a policy network. A policy network will develop in relation to a policy or set of related policies when political actors exchange resources regularly. The type of public policy and the intensity of the regulation which it brings about strongly affect the pattern of resource interdependency (Lowi, 1964, p. 688; Wilson, 1980, pp. 364-72; Daugbjerg, 1994, pp. 460-1). Any organisation entering the policy process is dependent on other organisations for resources and, consequently, they have to exchange resources in order to achieve their goals. As van Waarden (1992, p. 31) points out:

> [State actors] need political support, legitimacy, information, coalition partners in their competition with other ... [state actors], and assistance in the implementation of policy. Interest groups on the other hand desire access to public policy formulation and implementation of policy, and concessions in their interests or those of their members (see also Maloney et al., 1994, p. 36; Smith, 1993).

Policy networks emerge as a result of 'the dominance of organized actors in policy making, the overcrowded participation, the fragmentation of the state, the blurring of boundaries between public and private' (Kenis and Schneider, 1991, p. 41) and 'the increasing complexity of public affairs' (Campbell et al., 1989, p. 86). The modern state has a crucial role in economic and social life. To intervene in these areas, government needs resources which are not available within the state apparatus and cannot readily be developed (Kenis and Schneider, 1991, p. 41). It must, therefore, look elsewhere for such resources. Typically, it will become dependent upon organised interests which have resources within specific policy areas. This has resulted in policy making becoming complex, specialised and broken down into sub-systems (Richardson and Jordan, 1979, p. 43). In general, the function of policy networks is restricted to specific policy fields or sectors (Marin and Mayntz, 1991, p. 14), such as agriculture, health, labour market policies and energy, or sub-sectors within these broad policy areas. Furthermore, the network tradition ignores traditional policy making arenas, such as parliaments and political parties. It argues that these have only limited importance in sectoral policy making and that understanding the relationships involved in policy networks 'perhaps better accounts for policy outcomes than do examinations of party stances, of manifestos or of parliamentary influence' (Richardson and Jordan, 1979, p. 74).

A major concern within the network approach is to examine who is involved in public policy making and how actors, who form the core of a network, exclude other interests (Atkinson and Coleman, 1992, p. 159). Perhaps the most important reason why some actors are excluded from a network is that they lack valued resources. As mentioned above, exchange of resources is what characterises the policy process within a network. An actor with few resources to exchange can hardly achieve a central position because no organisations willingly include others. As Rhodes (1981, p. 122) elegantly points out: 'organisations are ... primarily concerned to avoid each other' (see also Laumann and Knoke, 1987, p. 13; Nordlinger, 1988, p. 882; Christensen and Christiansen, 1992, p. 7). Exclusion of some interests is a dominant feature of tight and closed policy networks. In these, members share common interests and views on policy problems (e.g. Jordan, 1990a, p. 327; Rhodes and Marsh, 1992a; Smith; 1993). This enables them to gain control over agenda setting,[5] policy formulation and implementation (Kenis and Schneider, 1991, p. 41).

Another common concern of the network literature is to analyse why policy processes and outcomes within many policy areas have remained stable over relatively long periods. To a large extent, stability can be explained by the presence of policy networks with restricted membership. Members of such networks have managed to depoliticise policy making by emphasising problem-solving based on technical rather than distributional discussions. They have, so to speak, 'technicised' the policy process. Parliaments, the public and the mass media have, therefore, found it very troublesome to intervene in the issue area because it is very difficult for them to acquire the skills which would enable them to discuss problems as equals. In other words, they do not have the resources which network members need. Insiders refuse outsiders access to the policy process by claiming that they are incompetent in the policy field (Baumgartner and Jones, 1993, p. 6).

Although we can speak of a policy network tradition, there are different approaches within it. As Marsh and Smith (1995, p. 1) point out: 'The problem is that whilst people agree on using the term policy networks, there is little agreement on what a policy network is or, perhaps more importantly, what a policy network does.' In the final analysis, the disagreement relates to one of the basic problems in political science, sociology and other social sciences: the structure-agency debate. In the extreme, the question in this debate is:

> Are ... actors unwitting products of their context, helpless individuals with minimal control over their destiny, floundering around in a maelstrom of turbulent currents; or are they knowledgeable and intentional subjects with complete control over the settings which frame their actions? (Hay, 1995, p. 189; see also Rothstein, 1988, p. 29).

In practice, most researchers position themselves somewhere in between these two poles, but in the final analysis, the debate comes down to whether we should attribute most explanatory force to structures or to the intentional actions of agents. The position empashising the intentional actions of agents does, however, observe the importance of the agent because structures do, in fact, leave some room for intentional action. The other position acknowledges that intentional agents are more or less constrained by the structure within which they act.

As in most other fields of political science, the structure-agency debate has also found its way into the policy network literature; however, the debate is not explicit but embedded in discussions on how to deal with the network concept (Marsh and Smith, 1995, p. 5). Four broad approaches can be identified in the policy network literature. These can be divided into two groupings, depending on the extent to which they emphasise structure or agency as the prevailing feature of policy networks. In the group emphasising agency, we find what Marsh and Smith (1995, pp. 6-8) call 'the group interaction approach' which is particularly associated with the work of Jordan, Richardson and their associates. The work associated with Wright and Wilks (1987; Wright, 1988) is also relavant here. However, highlighting the agency aspect of policy networks does not mean that the structural aspect is totally overlooked; it is just not given much attention.

The second group of approaches pays more attention to structure. Researchers favouring the formal network approach (e.g. Laumann and Knoke, 1987) are entirely concerned with structure and thereby overlook the importance of the agency. The approach associated with the works of particularly Marsh, Rhodes and Smith (e.g. Rhodes and Marsh, 1992a; Smith, 1993) underlines the importance of structure but also emphasises the role of agents.

The group interaction approach

The group interaction approach is particularly associated with the works of Jordan, Richardson and their associates. In the late 1970s, the approach attempted to develop concepts which could characterise entire political systems. In this light, Richardson and Jordan (1979) argued that policy communities characterised British politics. Although there was no link between the American literature on sub-systems/ sub-governments and the British literature on policy communities, they share some features (Jordan, 1981, p. 107; 1990a, p. 325). Both recognise that policy making takes place in relatively closed and specialised organisational settings; however, the policy community literature sees the policy process as more open than the sub-system and sub-government literature. The policy community literature emphasises that the number of participants in the policy process is much larger and more diverse than sub-system and sub-government writers suggest (Jordan, 1981, pp. 105-7). However, there is no return to classical or naive pluralism (Richardson and Jordan, 1979; Jordan and Schubert, 1990, p. 8). In the Nordic countries, political scientists applied a similar approach, corporate pluralism, in the late 1970s and early 1980s (e.g. Buksti and Johansen, 1979; Damgaard and Eliassen, 1978; Damgaard,

1981; Christensen and Egebjerg, 1979).[6] However, it was not internally coherent (Jordan, 1984, p. 146) and was incapable of doing more than rejecting corporatism (Heisler, 1979, pp. 281, 287-91;) and therefore never gained a foothold (Christiansen, 1994, p. 306). Indeed, corporate pluralism was no more than 'a description of the problem rather than a theory of causation' (Jordan, 1984, p. 147).

In the group interaction approach, a policy network is seen as a generic label for 'the ties between the bodies relevant to a policy area' (Jordan 1990a, p. 333; see also Jordan and Schubert, 1992, p. 10). The central concept of the approach is a policy community. It is 'a special type of stable network which has advantages in encouraging bargaining in policy resolution [and] ... exists where there are effective shared 'community' views on the problem. Where there are no such shared attitudes no policy community exists' (Jordan, 1990a, p. 327). Policy communities can be contrasted with issue networks. Following Heclo (1978), Jordan and Schubert (1992, p. 13) define an issue network as 'a relatively ad hoc policy making structure in which a large, and to an extent unpredictable, number of conflicting interests participate'. An issue network describes 'the politics of the ad hoc and the irregular' (Jordan, 1990a, p. 329).

The complexity of modern governance and the desire of civil servants and interest groups to create stability, certainty and predictability in their environments are the driving forces in the organisation of policy making. The creation of policy communities is the means for the fulfilment of these objectives. Complex situations and broad policy issues which involve the risk of being politicised are divided into manageable sub-issues. Policy making then becomes specialised and left to a few participants who have special interests in the technical details which characterises a sub-issue. By dividing an issue into sub-issues, policy makers can limit the number and range of groups who find their interests sufficiently affected to enter the policy process (Jordan and Maloney, 1995, p. 21). The result of specialisation is the emergence of many policy communities, each controlling policy making within narrow issue areas. Therefore, '[t]he politics of the policy community is the *politics of the particular* - a means to resolve the detail (and sometimes the substance) of politicized issues' (ibid., p. 19).

Within the approach, it is argued that a policy community's control over a policy issue rarely lasts for a long period. Closure and exclusiveness, which characterise a policy community, lead to political unrest among outsiders, eventually leading to politicisation of the policy field and new actors with resources may enter the policy making process. In the group interaction approach, the entry barriers are seen as being limited (ibid., pp. 17, 20) because, politically, no civil servant or government can afford to exclude a resourceful group (Maloney et al., 1994, pp. 23, 30). When a policy issue becomes politicised, the old policy community is broken up and an issue network emerges. This, in turn, creates an unstable and unpredictable situation within the policy sector. Old core actors will try to stabilise the situation by breaking the policy issue down into bargainable sub-issues. As a result, several, new, narrow policy communities will emerge (Maloney and Richardson, 1994, p. 125; Richardson et al. 1992, pp. 169, 173).

24

Jordan, Richardson and their associates view the drive towards fragmentation and specialisation as part of the logic of the modern policy process which implies that real policy making takes place in sub-sectoral policy communities. Sectoral policy communities cannot exist over long periods because policy issues dealt with at this level are too broad and involve too many diverse interests. Consequently, policy makers are forced to divide the policy issue into bargainable sub-issues which can be handled by sub-sectoral policy communities (Jordan and Maloney, 1995, pp. 22-23; Maloney and Richardson, 1994, p. 129; Jordan and Richardson, 1982, p. 88; Jordan, 1981; 1990, p. 333). This argument is challenged by adherents of the structural network approach who, in general, view sectoral networks as the more important in public policy making. The disagreement between the two approaches has focused on whether or not sectoral or sub-sectoral policy making characterises British agricultural politics. In the early 1980s, Jordan (1981, p. 116) argued that the agricultural policy process was characterised by sub-sectoral policy making in which a large number of interests had access to the policy process. Smith (1992, p. 28) argues that Jordan makes a methodological error by believing that presence in a policy arena is equivalent to having power. Although many groups may be involved in agricultural policy making, only a few have influence. When groups not belonging to the core of the sectoral policy community participate in the policy process, they do it within the institutional and ideological framework established by the core actors. In British agricultural politics, the Ministry of Agriculture and the National Farmers' Union are the core actors (ibid., pp. 31-32; see also Smith, 1990a). Smith, therefore, argues that it makes analytical sense to view agricultural policy making as sectoral (see also Cavanagh et al., 1995). Applying the same methodology as Jordan (1981), Jordan, Maloney and McLaughlin (1994) strike back, maintaining that agricultural policy making in Britain is fragmented and competitive and, thus, characterised by sub-sectoral policy communities. Defending their position, Jordan and Maloney (1995a, p. 631) claim: 'In our view the overall [agricultural] policy regime is the result of outcomes in specialist niches. There is not a dominant core which allows the specialists to negotiate within limits.' In an attempt to conclude the debate, Cavanagh, Marsh and Smith (1995, p. 627) argue that only empirical analysis can establish whether sectoral or sub-sectoral policy networks are the more important in a policy sector. However, since the debate involves major methodological disagreements, it is not yet concluded.

The analytical emphasis on sub-sectoral policy communities in the group interaction approach results from the limited role attributed to structures. The approach concentrates on the properties of agents and their interests in explaining policy making. What really counts when government officials decide whether or not an interest group should be incorporated into the policy making process are the resources it possesses, not its political goals and strategies (Maloney et al., 1994, pp. 21, 23, 28-29). Therefore, Maloney, Jordan and McLaughlin (1994, p. 30) argue that: '[t]he literature ... over-emphasises the development of norms of behaviour as the key variable in gaining legitimacy. We doubt whether a civil servant would ignore a resource rich group because it had behaved irresponsibly in the past.' Researchers working within the group interaction approach do not perceive the policy process as being significantly structured by principles, procedures and norms laid down by members of sectoral policy networks. They argue that 'real' policy making takes place at the sub-sectoral level. Quoting Schlozman and Tierney (1986, p. 396), Jordan and Maloney (1995, p. 18), therefore, argue that the most influential groups are those with narrow and technical goals rather than those with broader interests.

Although the group interaction approach attempts to downgrade the influence of structures, we cannot avoid structure in explanation building. In a manner which partly contradicts the argument that resources are the main driving force in determining membership of policy communities, Maloney, Jordan and McLaughlin (1994, p. 36) argue that 'more importantly, [groups] must also have "appropriate" goals'. In a similar vein, Jordan and Maloney, (1995, p. 22) suggest that 'a policy community can be seen as a mechanism for the assimilation of 'legitimate' competing values and the exclusion of those competing values deemed "illegitimate"'. In other words, structure influences participation. Research should, therefore, pay more attention to structure, in particular to the principles, procedures and norms guiding the policy process within sub-sectoral policy communities. These are often set by sectoral policy networks (Cavanagh et al, 1995, pp. 627-8). Therefore, understanding the policy process within sub-sectoral policy communities often (but not always) requires the researcher to examine sectoral networks.

The focus on sub-sectoral policy making and the reluctance to attribute explanatory force to structures are the main reasons why the group interaction approach has little to say about the choice of policy. At the sub-sectoral level, public policy making is very detailed and technical. Participants' policy positions are influenced by the attractiveness of specific and detailed technical solutions. It is difficult to find order in the disorder. The participants' policy positions seem to be unpredictable, even in the short term, and bargaining is influenced mainly by the search for technical solutions rather than by broader values. However, there will often be some sort of order in the empirical complexity. Broader concerns which are more or less explicitly expressed in policy principles structure policy outcomes. Focusing on these principles will enable us to predict policy outcomes, at least to some extent.

Applying the policy network as a metaphor (or label) makes the group interaction approach incapable of providing explanation, whether we are concerned with network transformation (Dowding, 1995, p. 139) or choice of policy. The reluctance to pay attention to structures implies that the approach views the public policy process as being fairly open. Thus, the approach is within the pluralist tradition (Smith, 1990, pp. 311-5; Jordan, 1981).

The personal interaction approach

The personal interaction approach is based on the assumption that the policy process is highly compartmentalised. It has developed a framework of analysis which is designed to examine policy making at a disaggregated level. In the personal interaction approach, networks are dependent, not independent, variables explaining policy outcomes.

Wilks and Wright distinguish between policy communities and policy networks. A 'policy community identifies those actors or potential actors ... who share a common identity or interest' (Wright, 1988, p. 606; see also Wilks and Wright, 1987, p. 298). A policy network is 'the linking process within a policy community or between two or more policy communities' (Wilks and Wright, 1987, p. 298; see also Wright, 1988, p. 606). In most of the network literature, 'the assumption has been made that a network is a general description of the ties between the bodies relevant to a policy area. The policy community is a certain type of network' (Jordan, 1990a, p. 333, see also Jordan and Schubert, 1992, p. 10). Wilks and Wright's approach uses the policy community and network concepts in a way which is at odds with the rest of the literature (Rhodes and Marsh, 1992, p. 17) and, therefore, contributes to confusion. As Jordan (1990a, p. 335) puts it: 'The main argument against the Wilks/Wright terminological usage is that they have been pre-empted. The terms they use already have an accepted currency and it is simply too confusing to use the terms differently.' Wilks and Wright (1987, p. 301) claim that this unique distinction between policy communities and policy networks 'enables us to identify those members of a policy community who are excluded from a policy network.' However, they do not argue convincingly that these properties are missing in other network approaches. Revealing who is included and excluded is essentially what the traditional policy community literature is all about (Jordan, 1990a, p. 335).

The main purpose of Wilks and Wright's framework is to facilitate the description of complicated policy processes, not to develop theory in the sense of reaching conclusions on causation (Wilks, 1989, p. 330). The limitations of the framework are obvious; nowhere in the articles in which Wilks and Wright develop the framework, do they suggest any possible links between network and outcomes. This inability to explain outcomes is the result of their desire to disaggregate to what they call the sub-sectoral level but which is, in fact, the sub-sub-sectoral level. Wright (1988, p. 597) seems to suggest that educational policy making should be disaggregated to policy making in higher, further, secondary, primary and nursery education. At that level, inter-personal rather than inter-organisational relationships have an important impact on policy outcomes. Wilks and Wright do not explicitly say that in a network, inter-personal relationships are more important than structural relationships; implicitly, however, they place 'considerable emphasis on interpersonal relations as key aspects of the policy networks' (Rhodes and Marsh, 1992, p. 17; see also Rhodes, 1990, p. 307). This emphasis is clear in their review of the case and comparative studies in the book *Comparative Government-Industry Relation* (Wilks and Wright, 1987a). They conclude that '... we have a series of cases where government-industry relations are mediated through quite distinct personalized or informal sets of relationships ...' (Wilks and Wright, 1987, p. 287). Emphasising the inter-personal relations of policy networks implies that the network itself has only a very limited role in explaining policy outcomes. What counts for Wilks and Wright are the resources which individuals who are members of a network possess. The network concept is, in fact, superfluous in the personal interaction approach if one

27

wants to explain policy outcomes. Explanations obtained by using their approach could be achieved by the use of other approaches, in particular rational choice institutionalism.

The formal network approach

In the formal network approach, the network itself to a large extent explains the actions of individuals and organisations. As Knoke (1990, p. 7) argues: 'The basic units of any complex political system are not individuals, but positions or roles occupied by social actors and the relations or connections between these positions' (see also Laumann and Knoke, 1987, p. 9).

In analysing policy networks, the primary analytical focus 'is on the relational connections as such - the ties among the positions - and not on the attributes of the incumbent individuals who occupy these positions' (Knoke, 1990, p. 8; see also Laumann and Knoke, 1987, p. 226). The relations among members of a network and how they are positioned in the network structure have an important influence on their perceptions, orientations and actions. Consequently, networks facilitate some political actors' participation in policy making and constrain others' (Laumann and Knoke, 1987, p. 226; Schneider, 1992, p. 110; Knoke, 1990, p. 9).

The formal approach has not yet developed into a network theory which can explain policy outcomes. The approach is mainly a collection of analytical concepts which can be used to produce propositions on participation in policy making. Actors' location within the network structure is the basic analytical concept. Location has important consequences for their participation and influence. Actors centrally located in communication and resource exchange networks participate more in policy events[7] than actors located at the periphery (Laumann and Knoke, 1987; Knoke, 1990, pp. 11-16; Laumann et al., 1985). Those who receive information from many others and have short lines of communication are those who are located most centrally (Knoke, 1990, p. 10; Laumann and Knoke, 1987, p. 13). Actors' location in resource exchange networks is determined by their ability to gain control over scarce resources and minimise the dependence on other actors' resources (Knoke, 1990, p. 13). A major purpose of the communicative process within a network is to frame a policy problem so that it both encourages and discourages actors from entering the policy process. While centrally positioned actors have significant advantages in framing a debate, outsiders or peripheral actors face difficulties in having their views accepted (Laumann and Knoke, 1987, pp. 315, 322).

The formal network approach applies highly sophisticated statistical methods to map actors' positions in a network. To use such methods, researchers need quantified data. The sole reliance on quantitative data and statistical methods is the most significant weakness of the approach. These constraints in research limit the approach's ability to discover the quality of relations among actors and reach conclusions which go beyond what we already know (see Marsh and Smith, 1995, p. 13; Atkinson and Coleman, 1992, pp. 160-1).[8] For instance, Laumann and Knoke (1987, p. 284) are only able to conclude: 'At present, we can be fairly confident in concluding that when an organization believes that its interests are at stake, it will attempt to influence the direction of such a decision regardless of the resources available for that effort' (see also Laumann et al., 1985, p. 17). Such a conclusion is disappointing relative to the resources (money and time) needed to use the approach.

The major problem of the formal network approach is its inability sufficiently to account for the quality of relations between actors. It is not enough to characterise networks in terms of 'their level of integration, their degree of openness or even by the coalitions that have formed around certain policy options' (Atkinson and Coleman, 1992, p. 161). In a similar vein, Marsh and Smith (1995, p. 14) point out: 'Having a lot of contacts in a network does not mean that a group is influential ... Mapping networks solely using quantitative data tell nothing about the quality of interaction and even less about the degree of influence.' Network analysis needs also to focus on institutional and ideological variables (Atkinson and Coleman, 1992, p. 161). Statistical analysis does not have much to offer in demonstrating the influence of these.

The strength of the formal network approach is that it demonstrates that network structures play important roles in government-interest group relations. What is more, it provides an advanced research method to identify policy networks. However, the identification of networks will not in itself enable us to predict policy outcomes (ibid., p. 161; Marsh and Smith, 1995, pp. 13-14). This is perhaps the reason Laumann and Knoke (1987) avoid asking questions about policy outcomes and limit themselves to examining who participates in the policy making process. The inclusion of institutional and ideological variables and the application of other research methods would provide better opportunities to answer the questions not asked by the two scholars.

The structural approach

Compared to the formal network approach, the structural approach is much more modest in its demand for resources and is also able to provide more interesting findings, even though it uses less sophisticated research methods. In agreement with most of the network literature, the structural approach uses the term 'policy network' as a generic term which encompasses all types of government-interest group relationships. The structural approach is a further development of the Rhodes model for the study of central-local government relations (Rhodes and Marsh, 1992a, p. 187). In the Rhodes model, the policy network concept was used for the study of relationships between central and regional/local authorities. The structural approach extends network analysis to the relations between government and organised interests. Since this approach draws heavily on the Rhodes model, it is useful to describe the latter briefly.

The Rhodes model explicitly distinguishes between different types of policy networks (policy communities, territorial communities, issue networks, professionalised networks, inter-governmental networks and producer networks). The type of network which emerges depends on the interests and actors represented in the network, the degree of both vertical interdependence (intra-network relationships) and horizontal interdependence (inter-network relationships) and the pattern of resource distribution within and between networks (Rhodes, 1986, pp. 22-23; 1992, pp. 77-78).

By explicitly emphasising variation among policy networks, the Rhodes model captures an essential aspect of meso-level policy making. Whereas the group interaction approach pays more attention to common features of meso-level policy processes rather than the variation among sectors (see Smith, 1993, p. 57), Rhodes realises that there is too much variation to justify a claim of common features across policy sectors. Theoretically, however, the Rhodes model is weak in its attempts to explain the ways in which policy networks influence policy outcomes. Empirically, Rhodes (1992) observes that tight and closed networks (policy communities) tend to produce stable policy outcomes, whereas loose networks tend to be associated with unstable policy outcomes. Although his work is theoretically weak in terms of explaining policy outcomes, he argues 'that the analysis of policy networks cannot be limited to an analysis of process; it must encompass policy content' (ibid., p. 83).

A major reason why Rhodes is unable to establish the theoretical link between network type and policy content is that he pays too little attention to network structures. The pattern of resource exchanges in policy networks is given too much emphasis at the expense of the structural constraints and opportunities within different types of networks. This insufficient focus on structure results from the view that interaction between network members is a game in which 'both sides manoeuvre for advantage, deploying the resources they control to maximize their influence over outcomes, and trying to avoid (where they can) becoming dependent on the other "player"' (Rhodes, 1986, p. 18; see also 1992, p. 42). The constraints and opportunities embedded in a network influence the actions of its members. Explanation, therefore, demands more attention to structure. Another problem of the Rhodes model is the use of fixed network types. By applying only six types, Rhodes leaves the impression that there are no more. In reality, policy networks take many forms - many more than six. In this respect, the model is too rigid.

Marsh and Rhodes attempt to overcome some problems of the Rhodes model. Their approach is particularly designed for the study of relationships between state and societal organisations when they make public policy. Resource interdependency remains at the centre of the approach. Instead of operating with fixed categories of policy networks, Marsh and Rhodes apply a continuum of networks based on the application of four analytical dimensions: membership, integration, resources and power. Policy communities and issue networks are the two end points on a continuum with many intermediate types of networks.

Table 1.1

Marsh and Rhodes' policy network typology[9]

Dimension	Policy community	Issue network
Membership Number of participants	Very limited number, some groups consciously excluded	Large
Type of interest	Economic and/or professional interests dominate	Encompasses range of affected interests
Integration Frequency of interaction	Frequent, high-quality, interaction of all groups on all matters related to policy issue	Contacts fluctuate in frequency and intensity
Continuity	Membership, values and outcomes persistent over time	Access fluctuates significantly
Consensus	All participants share basic values and accept the legitimacy of the outcome	A measure of agreement exists, but conflict is ever present
Resources Distribution of resources (within network)	All participants have resources; basic relationship is an exchange relationship	Some participants may have resources, but they are limited, and basic relationship is consultative
Distribution of resources (within participating organizations)	Hierarchical; leaders can deliver members	Varied and variable distribution and capacity to regulate members
Power	There is a balance of power among members. Although one group may dominate, it must be a positive-sum game if community is to persist	Unequal powers, reflecting unequal resources and unequal access. It is a zero-sum game

Marsh and Rhodes emphasise variation, stating that 'the policy network approach is based on the assumption that any adequate characterisation of the policy process must recognise variety' (Rhodes and Marsh, 1992a, p. 188). This observation leads the approach towards explaining the emergence of different network types (Thomsen, 1996, p. 27). Smith in particular is concerned with this question. Building

31

on the structural approach, he argues that policy communities emerge in some policy areas because interest groups and state actors involved have considerable resources to exchange. Groups exchange 'information, legitimacy [and] implementation resources ... for a position in the policy process, and some control over policy' (Smith, 1993, p. 63). In an issue network, actors have only limited resources to exchange and, therefore, the exchange process is less advanced. As Smith (1993, p. 63) points out:

> Most of the interest groups are likely to have little information to exchange and little control over the implementation of policy. Consequently, they are forced into overt lobbying activities. Government agencies, which do not have a monopoly of the policy area, cannot guarantee that an interest group has a role in the policy process.

The ability of interest groups and state actors to gain control over resources influences the type of policy network which emerges. The membership of a policy community is limited to one or few interest groups. If there is more than one, they are rarely competing to organise the same social group (ibid., pp. 59-60). State actors also need to gain control over resources, primarily authority, if a policy community is to develop. Conflict between government departments or agencies as to who is responsible for a policy area is often the most important reason a policy community does not develop. Government departments or agencies in conflict will, in order to strengthen their position, try to draw supporting interest groups into the network. As a result, a consensus within the network is unlikely to emerge. The relationship between actors tends to develop into an issue network which has many participants and suffers from frequent and fundamental conflicts over policy making (ibid, p. 63).

The structural approach is also concerned with the way in which different types of policy networks influence policy outcomes. Unfortunately, the approach has not moved much beyond the observation of 'correlations.' There are only few studies which account for the causal links between policy networks and policy outcomes, although the approach holds that we should give emphasis to structures in explaining policy making. Marsh and Rhodes (1992, p. 262) suggest: 'The existence of a policy network, or more particularly a policy community, constrains the policy agenda and shapes the policy outcomes. Policy communities, in particular, are associated with policy continuity' (Rhodes and Marsh, 1992a, p. 197). Therefore, in policy areas in which a policy community exists, we cannot expect more than incremental policy change, unless all its members agree to bring about basic change. Despite the notion of causation, there are only few contributions within the structural approach which actually try to explain policy outcomes. In the case study collection *Policy Networks in British Government* (Marsh and Rhodes, 1992a), the contributors have little to say about the relationship between policy networks and policy outcomes. They merely point to the importance of network structures. Smith (1992, p. 48), for instance, argues that 'structures and ideologies ... [of a policy community] ... determine the options available and the groups involved within the policy community.' In a similar vein, Wistow (1992, p. 52) limits himself to arguing that 'the values and interests embedded within [a] network determine the distributional outcomes of policy making'. In drawing conclusions on British youth employment policy, Marsh (1992, p. 198) is also very modest, merely proposing that a causal relationship between policy networks and policy outcomes exists. Stones (1992, p. 224) and Peterson (1992, p. 229) argue briefly in a similar way. Saward (1992), Read (1992), Cunningham (1992) and Mills (1992) have nothing to say which reaches beyond the empirical cases investigated.

Martin Smith (1992, 1993) is the analyst within the structural approach who is most concerned with the influence policy networks exercise on policy outcomes. He argues that the keys to understanding policy outcomes are the ideologies and institutional structures of networks. Although Smith makes an important contribution to the policy network literature, he does not, however, develop theoretical models which establish the relationship between network type and policy type. He merely indicates the importance of structures in policy making by showing the 'correlation' between the type of network and a broad, and not very precise, conceptualisation of policy outcomes. Smith points out that a policy community's institutional and ideological structure limits the range of available solutions: 'Policy is limited to what is acceptable to the consensus within the policy community. The policy community is effectively a means of excluding certain policy options' (1993, p. 71). In issue networks, by contrast, 'policy outcomes are likely to be more varied ... [because] ... it is much easier for groups to get alternative policy options onto the agenda (ibid.). Smith's arguments need further development in order to provide a theoretical model which links various policy network types with policy choices.

Smith argues that a policy network's ideology has an important influence on the choice of policy. He says that in a policy community: 'Ideology defines not only what policy options are available but what problems exist. In other words, it defines the agenda of issues with which the policy community has to deal. Therefore members of a policy community agree on both the range of the existing problems and the potential solutions to these problems' (1993, p. 62). However, although he identifies a crucial factor influencing the choice of policy, he does not make sufficiently clear the way in which ideology relates to policy outcomes. Firstly, his analysis remains at a too general level and does not consider the way in which ideology defines some policy solutions as attractive. Consequently, he cannot precisely pinpoint how ideology affects the contents of public policy. Secondly, and perhaps most importantly, Smith does not sufficiently consider what constitutes a public policy. Like other network analysts, Smith applies the concept of policy outcome in a way which is far too loose, broad and imprecise; he only classifies policy outcomes in terms of its stability or instability over time. In order to establish causal links between policy networks and policy choices, much more attention must be paid to the contents of policy, and classifications which fit the logic of policy making within various network types must be developed. We can only establish why some interests benefit from policy outcomes and others lose if we understand policy choices. Thus, a theoretical model establishing the links between policy networks and policy choices requires a clear definition and a precise classification of policy. The policy network literature has not paid enough attention to this problem.

Network analysis and policy making in the European Union

More or less explicitly, the network concept underpins several studies of European Union (EU) policy processes (e.g. Bretherton and Sperling, 1996; Grant, 1991; 1993a; Mazey and Richardson, 1992; 1992a; 1993, 1993a; Peterson, 1992; Smith 1990, ch. 6, 7; Daugbjerg, 1994). Although they employ the network concept, none of these studies explicitly questions whether it makes analytical sense to apply it at the European Union level. Essentially, policy network analysis is developed to analyse national policy processes. This section assesses whether or not it can be applied at EU level.

There is no doubt that the EU policy process is different from policy processes in most European nation states. As Mazey and Richardson (1993, p. 206) point out:

> ... the distinctive nature of EC policy-making structures and processes - notably the openness of the decision-making process, its multinational character, and the considerable weight of national politico-administrative élites within this process - create an unstable, multi-dimensional environment within which groups have to operate.

In particular, the process of agenda setting at EU level is much more open than in nation states (Peters, 1994, p. 11). In the EU, 'new ideas and proposals can emerge from nowhere with little or no warning' (Mazey and Richardson, 1993a, p. 22), resulting in unstable and unpredictable agendas (ibid., p. 11; 1992a, p. 110). However, despite the differences in policy making environments, national and EU lobbying have some common features. Resource interdependency is an important variable explaining the exclusion and inclusion of political actors. Additionally, informal contacts play an important role in the relationships between public and private actors in both settings (Mazey and Richardson, 1993, p. 206).

Mazey and Richardson (1993b, p. 253) argue that at EU level, 'the network concept is quite useful' but they do not 'define their terms or explain why the concept is useful' (Peterson, 1995a, pp. 390-1). Peterson (1995) argues that network analysis is particularly suited to studies of EU policy processes mainly for two reasons. Firstly, since the European Union is a relatively young political system it lacks formal and well-established institutions (p. 82) 'which can facilitate bargaining between different types of actors at the meso-level' (p. 71). To cope with the uncertainty resulting from the lack of formal institutions which can mediate between interests involved in EU policy making, EU policy makers have set up policy networks to achieve a minimum of predictability and stability in the policy process (p. 82). Secondly, Peterson argues that many important EU policy decisions are effectively made in the early phases of the process (p. 86; 1995a, p. 397). Since the

Commission has the sole right to initiate policy proposals, it is often a crucial political institution in these early phases. This gives Commission officials an important role in preparing proposals. In general, however, they depend on resources and advice from outside. Compared to national bureaucracies, the EU bureaucracy is relatively small and, therefore, relies on external sources of information and knowledge (e.g. Mazey and Richardson, 1992a, p. 115; 1993, p. 209). In its relations with the Council of Ministers, it depends on support from interests groups, in particular European associations. The Commission knows that 'any proposals that are opposed by affected interest groups are far less likely to survive the scrutiny of the Council of Ministers' (Butt Philip, 1985, p. 44). On the other hand, interest groups depend on the Commission to further special interests or to prevent its proposals from violating such interests. In order to promote or protect national interests, national civil servants and interest groups also need the Commission. Because these different groups of actors depend on each other's resources, policy networks develop. These can be tight policy communities or loose issue networks.

However, there are some difficulties in applying the network concept at the EU level, although these are not as great as Kassim (1994) suggests. He claims that the concept has little to offer because the EU policy process is too changeable and fragmented. Furthermore, he argues that it is not sufficiently sensitive to the institutional complexity of the European Union and is not equipped to delineate networks at EU level. Clearly, Kassim exaggerates the problems. Firstly, the application of the network concept does not depend upon stable and centralised decision making processes; these features are not assumed in network analysis. Rather, it treats these features as variables. Fragmented and changeable policy processes occur because issue networks prevail within an issue area. Secondly, network analysis does not assume that the presence of a single decision making centre within a policy area is a condition for the development of policy networks. It treats institutional complexity as a variable, not as an assumption (Peterson, 1995a). In this way, competing decision making centres within a policy area are the most important reason for the development of issue networks (Smith, 1993, p. 63). Finally, the delineation of the boundaries of EU policy networks is not impossible although it requires a greater effort than at the national level because the number of actors with resources is greater (Peterson, 1995a, pp. 402-3).

Although Peterson (1995a) rejects most of Kassim's criticisms, there are reservations about the utility of network analysis at EU level. In its explanation of the contents of EU policies, network analysis has an important limitation. It requires that, in practice, policies are made in the early phases of the policy process. Peterson (1995, p. 86) is one of the EU analysts who argue that this is often the case. As he points out: 'More than in most national systems, eventual outcomes are shaped in crucial ways early in the policy-making process, especially when different actors bargain and exchange resources as policies are formulated.' However, we do not yet know enough about EU processes to treat that argument as an assumption. The question remains an empirical one. If EU networks involving Directorate Generals, national government departments or agencies and organised interests cannot reach agreement in the early phases of the policy process because the issues they deal with are politically sensitive, then the Council of Ministers, which has the right to make binding decisions, becomes the major decision making body. In such cases, the policy process involves intense negotiation and bargaining among national governments. In other words, inter-governmentalism characterises the process. In such situations, network analysis at EU level has only limited value. In order to understand EU processes characterised by inter-governmentalism, national policy networks should be the centre of the analysis because they have an important influence on member states' policy positions. By contrast, EU networks become crucial determinants of policy when the Council of Ministers accept agreements reached in such networks and, therefore, adopts them. In order to use network analysis at the EU level, we need to achieve a better understanding of the EU policy process; that is, to know whether and when EU or national policy networks become important determinants of policy. Although network analysis has limitations in accounting for final EU decisions, it does help to explain the Commission's proposals. As already mentioned, the Commission is a relatively small bureaucracy which has to rely on information from outside sources, and depends on the support of interest groups in its attempts to persuade the Council to approve its proposals. This need for information and support is an important reason EU policy networks emerge. It is important to understand these networks because they affect, in varying degrees, the content of the Commission's proposals which are the basis of the decision making process within the Council of Ministers.

This book uses network analysis at the level of the European Union in order to understand the Commission's position in the 1992 agricultural policy reform. The application of network analysis does not involve, therefore, insurmountable difficulties.

Can the policy network approach move from description to explanation?

In the network approaches emphasising structure (the formal approach and the structural approach), policy networks become more than a metaphor. The characteristics of the network have an important influence on policy outcomes. However, the structure of the network does not in itself determine outcomes (Rhodes and Marsh 1992, p. 2; Marsh, 1995, p. 4). Applying the network concept as more than a metaphor (or a label) means that it becomes analytically meaningful. A network theory can be developed if we conceive of '[p]olicy networks ... as *specific structural arrangements* in policy making' (Kenis and Schneider, 1991, p. 41). Using the concept solely as a metaphor does not, however, help to provide an explanation of policy making.

The review of the four network approaches clearly shows that their current strength is in description rather than in explanation. However, the structural approach has potential to develop theoretical explanations. It emphasises the importance of the network structure as a factor which can contribute to an explanation of policy outcomes; the approach does, however, need more development in order to achieve this aim. In particular, we need to construct classifications of policy outcomes which can be linked to the policy process within different network types. The chapter shows that policy network analysis can be used for the study of European Union policy processes although such a usage has some limitations.

Policy networks influence the actions of both insiders and outsiders. They lead them in certain directions. Their members (insiders) interpret the context within which they are embedded and this interpretation then forms the basis of their actions. Outsiders also interpret the structure of networks. Most often, they perceive them as disadvantageous to their chances of gaining influence in the policy process. Consequently, they refrain from action. However, in some, but infrequent, situations, outsiders face an open 'policy window' (Kingdon, 1984) and therefore decide to enter the process.

Notes

1. See Jordan, (1990b) and Smith (1990) for a debate on pluralism.

2. See McFarland (1992), Jordan (1990a) and Rhodes and Marsh (1992) for an overview of the literature.

3. Later (in 1979) Schmitter (1979, p. 65) saw corporatism as a system of *interest intermediation* rather than a system of *interest representation*.

4. For an overview over corporatism, see Williamson (1989). See also Grant (ed.) (1985a), in particular Grant (1985).

5. However, agenda setting 'may pass over discussions in the public and in the mass media, but very often policy issues are raised and defined within restricted networks' (Kenis and Schneider, 1991, p. 41).

6. See *Scandinavian Political Studies, vol. 2, no. 3, 1970* for more examples and references.

7. Laumann and Knoke (1987, p. 30) define an event 'as a critical, temporally ordered decision point in a collective decision-making sequence that must occur in order for a policy option to be authoritatively selected.'

8. The methodological considerations of the approach are dominant. This leads Wolman (1995, p. 22) to suggest that 'one cannot avoid the impression that the techniques, rather than the theoretical concerns, are the driving force in the nature of the questions asked.'

9. Adapted from Rhodes and Marsh (1992a, p. 187) and Marsh and Rhodes (1992, p. 251).

2 Policy networks and policy changes: a meso-level analysis

Introduction

In chapter 1, it was concluded that the network literature is relatively weak in explaining policy outcomes. It has not yet developed into a theoretical model explaining the choice of policy (Mills and Saward, 1994, p. 87; Rhodes and Marsh, 1994, p. 7). However, network analysts have made some important observations: a policy community produces policy continuity, whereas loose networks are associated with unstable and unpredictable policy outcomes (e.g. Smith, 1993, pp. 71, 75; Rhodes and Marsh, 1992a, p. 197; Marsh, 1995). When it comes to explaining change, policy network analysis also needs further development (Atkinson and Coleman 1992, p. 172). As Rhodes and Marsh (1992, p. 15) point out: 'It might be argued that most of the literature on policy networks has paid insufficient attention to the question of change; certainly such a failing is not surprising given the emphasis on policy networks as a barrier to change.'

A major claim in network analysis is that the contents of public policies are associated with the shapes of policy networks. To give this argument force, it is necessary to develop a classification of policy networks which includes only those features of networks which have significant impact on the policy process. When developing network classifications, network analysts have not been careful to distinguish the structural features of networks from the properties of the agents or the characteristics of policy outcomes. The unclear distinction between these three aspects of network limits the ability of network analysis to explain policy outcomes. Therefore, before a network model explaining policy outcomes can be developed, one needs to develop a network continuum which operates with only structural features of the network itself. In other words, it must keep network structures, network members and policy outcomes separate.

Pressure for change can arise both from inside and outside policy networks. This book attempts to explain policy choices when actors attempt to bring about change in sectors in which they do not traditionally participate in policy making. These actors can be called outsiders. Changes may also be initiated within networks, but the theoretical model being developed here does not apply to such situations. Usually, outsiders' proposals have a negative image for the group which is subject to potential changes. Because outsiders find existing situations disadvantageous, they base policy proposals on principles which contradict established sectoral policy principles. Insiders view established policy principles in a positive light because they contribute to a highly valued way of life and because they provide social, political or economic benefits. Since insiders and outsiders have different views on policy principles, conflicts are likely to occur. For instance, in the 1980s and early 1990s, farmers viewed pollution control and policy reforms in a negative light because environmentalists and agricultural reformers based their policy proposals on principles which did not coincide with those already established. Environmental interests favoured the polluter pays principle and agricultural policy reformers based their reform proposals on market principles, i.e. the state should cease to protect farmers from the market. These principles contradicted the major policy principle of agricultural policies: the principle of state responsibility for economic imbalances. Therefore, agri-environmental policy making and agricultural reform processes involved a clash between conflicting policy principles (see the introductory chapter for an account of these conflicts).

The struggle over policy principles is an important, if not the most important, aspect of politics (Krasner, 1984, pp. 224-5; March and Olsen, 1989, p. 37; Baumgartner and Jones, 1993, p. 89). This position does not invalidate the traditional one, which sees politics as the authoritative allocation of values, whether 'material or spiritual' (Easton, 1953, ch. 5). The struggle over principles is the struggle for control over future authoritative allocation of values.

Originally, the network concept was not developed to explain change but it does have this potential. Whether outsiders succeed in bringing about unwelcome reforms of existing sectoral policies or in introducing new, unwelcome policies depend on the policy network existing in the sector concerned: 'Where there is an issue network or looser network the effect of social and political change is likely to be greater than where there is a closed policy community' (Smith, 1993, p. 87). This statement is merely an observation of the relationship between two variables. To give Smith's observation explanatory force, we need to specify which 'mechanisms' enable policy communities to limit outsiders' effects on policy to a much greater extent than looser networks. Besides influencing policy choices, outside pressure can also lead to network change. Whether or not established policy networks can survive outside pressure is not addressed in this book but is, nevertheless, an important question which should be examined in future research.

Policy change can be either fundamental or moderate. *Fundamental policy change* introduces new policy principles as the basis of policy choices. In agriculture, fundamental policy change takes place when market principles or the polluter pays principle, instead of the principle of state responsibility, is used to underpin the policy. *Moderate policy change* does not involve a shift in policy principles. Established policy principles continue to underpin the choice of policy. Existing policy instruments and objectives may be adjusted or new ones may be introduced, but the nature of the policy remains the same. A moderate policy change in agriculture may involve the use of new instruments in agricultural policy, but these do not contradict the principle of state responsibility.

The meso-level theoretical model which is developed in this chapter argues that the success of outsiders depends mainly on the policy network type existing in the sector in which they attempt to bring about policy change. If there is a high degree of cohesion among network members, interest groups may attract significant support from the other members, in particular state actors. Consequently, outsiders have to accept that fundamental policy change does not occur. On the other hand, if the degree of cohesion among members of the network is low, outsiders have a much greater chance of bringing about fundamental policy change. I recognise that in reality, many types of policy networks and policies exist. To avoid a very complicated model, I discuss only extreme situations; that is, I link cohesive and non-cohesive policy networks with policy change. However, in reality, most cases are in between these two extremes.

Dimensions on the policy network continuum: revising the Marsh and Rhodes continuum

Policy network analysts have developed various classifications of networks. In reviewing the literature on policy networks, one may easily become confused by the great variety of dimensions applied to define networks. The variation concerns both the number and the characteristics of the variables. Some network analysts apply few dimensions to describe networks, (see e.g. Grant, 1992, pp. 57-8; Jordan and Schubert, 1992, p. 12), whereas others have developed sophisticated classifications consisting of many dimensions (see e.g. van Waarden, 1992). A major problem within the literature is to select the most important dimensions. Choosing whether the network classification should consist of few or many dimensions involves a choice between parsimony and detailed description. The former is emphasised here.

The many different dimensions and variables used to develop network classifications and typologies are an obstacle to the development of a network theory. What is needed are criteria by which we can establish the variables which should define a policy network. Dowding is aware of this problem in that he draws attention to the unclear distinction between properties of networks and properties of their members. Correctly, Dowding (1995, p. 153) points out that the '*analytical* division into members' characteristics and network characteristics can enable us to keep in clearer view the relationship of variables in any given explanation.' In other words, one should keep structure and agency separate. In choosing the dimensions which best characterise policy networks, it is essential that one does not employ the outcomes of networks as properties of networks. Network analysts do not always meet these criteria. For instance, Marsh and Rhodes (1992, p. 251; Rhodes and Marsh, 1992a, pp. 186-7) include the distribution of resources within the network as well as within the participating organisations as characteristics of a policy network. But resources are properties of actors (Dowding, 1995, p. 153).[1] Similarly, Van Waarden (1992, p. 37) includes actor strategies as a property of a network but in his usage of the concept, it is a property of the agent. He is concerned with the way in which state agencies choose network building as a strategy to satisfy their needs, interests and goals.[2] Network analysts also use dimensions which are outcomes as characteristics of networks. For instance, Grant (1992, p. 57) identifies stability and internal cohesion as attributes of networks. Admittedly, I have also used stability to characterise networks (Daugbjerg, 1994, p. 460). More properly, both stability and internal cohesion are outcomes of policy networks. Networks also produce power relations and, as such, power is an outcome and, consequently, it should not be used as a network property.

The unclear distinction between the properties of actors, networks and outcomes limits the utility of the network concept in its current state. A more careful consideration of which properties of networks have most explanatory power will undoubtedly increase their value as an explanatory concept. In its present form the network concept is, first and foremost, 'a means of categorising the relationships that exist between groups and the government' (Smith, 1993, p. 56) or, in Rhodes and Marsh's (1994, pp. 7-8) words, it 'is a tool for analysing the policy process and intermediation and, like any tool, it only does the job for which it was designed.' The Marsh and Rhodes continuumdoes a satisfactory job in that respect. By operating with a continuum, Marsh and Rhodes have developed a tool which is superior to other network classifications operating with fixed categories.[3] The network continuum provides the researcher with flexibility in empirical research and, thus, seems much more fruitful than efforts to develop few fixed categories. The use of only few categories will make it difficult to place an empirical case into such categories without losing important information.

Moving the structural approach forward requires a continuum which stresses those properties of a network which have the most explanatory force; that is, those properties which have significant influence on the choices made by political actors. In chapter 1, a policy network was defined as a structural arrangement; thus a network continuum should describe the context within which the members act. In particular, membership, integration and institutionalisation have an important influence upon policy decisions.

Who does and does not participate has an important impact on policy making and, thus, membership is a significant feature of a network. It defines whose interests are the most important in policy making. Interests represented by actors outside the network are given low priority; however, they need not be totally neglected. In order to avoid mobilisation of opposition outside the network, members may, to some extent, take the interests of outsiders into account. The Marsh and Rhodes continuum operates with type of interest as a variable, suggesting that in a policy community economic and/or professional interests dominate. However, several studies show that a closed policy community can also consist of non-economic or non-professional interests (e.g. Svold, 1989, pp. 63-9). Therefore, range of interest seems a better variable. The interests the dominant state actor in a policy network represents also influence what type of network will emerge. Some state actors are concerned

41

with broader public interests or with counterbalancing a dominant interest group in the network. Other state actors 'choose to view themselves solely as representatives of constituent and client interests, thinking it appropriate and correct that they act solely according to the latter's preferences rather than their own' (Nordlinger, 1981, p. 37).

Integration is the second crucial dimension. It defines the form, quality and frequency of interaction within the network. Members of policy communities are highly integrated in governmental policy making whereas members of issue networks are only loosely integrated. Interaction can take a variety of forms, ranging from bargaining and negotiation in policy communities to consultations in issue networks (Rhodes and Marsh, 1992, p. 187). The use of bargaining as a mode of interaction presupposes that the network has conflict management procedures which define one or a few acceptable dimensions of controversies. Even in a tight policy community, there is rarely a consensus on solutions to all specific problems; however, there is likely to be a consensus on the way in which to reach agreement (Grant et al., 1988, p. 12), given that the controversies are limited to a few and well-known dimensions. Often, 'issues may be defined to include only one dimension of conflict' (Baumgartner and Jones, 1993, p. 19). For instance, until the mid-1980s, controversies in agricultural policy making involved only one dimension. It concerned 'farm subsidy benefits versus budgetary costs. Other dimensions of conflict, say environmental consequences ... would not intrude' (ibid.). If the controversies are within narrow, accepted boundaries, policy makers can settle them by using well-known methods of conflict management. Consultation, which prevails in an issue network, is non-binding and the extent to which state actors consult the other network members varies. No member is totally excluded from the policy process; all members are consulted. Usually, bargaining is impossible because there are no procedures for conflict management. Conflict evolves, therefore, along many dimensions.

Interaction also varies in frequency. Contacts are frequent and relate to all matters within the policy issue in policy communities (Rhodes and Marsh, 1992, p. 186) and, therefore, network members are 'constantly involved in the policy process' (Smith, 1993, p. 62). In issue networks, the pattern of interaction is unstable (Rhodes and Marsh, 1992, p. 187). Who needs whom varies from issue to issue and even from question to question. An interest group will only be consulted if it has resources which are relevant to the specific question on the agenda; in other situations, it will be marginalised.

Integration can be formal or informal. The former refers to interaction authorised by law or by other forms of official statements which are issued by the parliament, the cabinet or a minister. Informal integration can be defined as interaction which is not authorised by law or by other forms of official parliamentary, cabinet or ministerial statements, but is regarded as legitimate. In practice, integration is often a combination of both types.

Institutionalisation is the final dimension of the continuum. This aspect of policy networks is under-researched. As Atkinson and Coleman (1992, pp. 175-6) point out: 'analysts must seek to ascertain the more general principles and norms underlying interpretations of the policy field. They must observe the principles and norms over time and they should be alert to the possibility of competing beliefs within the given policy community.' The degree of institutionalisation can be defined by the extent to which there is a consensus on the principles to underpin policy choices and on the procedures with which to approach policy problems.

Policy principles define what should be the basic idea of public policies and outline, in general terms, 'what solutions are attractive and feasible' (Campbell et al., 1989, p. 86). The principles underpin the choice of policy. If network members have developed a consensus on the principles, then these come to serve as the network's code of good policies. Consequently, they are rarely discussed; they are only discussed when network members are forced to do so because ongoing discussions on policy principles would destabilise the network. Policy principles may remain stable for decades when members have reached a consensus on them.

In most policy sectors, the definition of the state's role in the economy is the very core of public policy making. Therefore, the most basic policy principle in a network dealing with public intervention in an economic sector concerns the role of the state; that is, 'what sort of activities ... [the state should] be involved in, and what form that involvement should take' (Cerny 1990, p. 51). At one extreme, the state has a mini-malist role which means that it is mainly engaged in establishing and maintaining competitive markets (ibid., p. 80). Its task is then to provide a framework helping market forces to operate effectively; that is, ensuring that competition is not dis-torted. A minimalist role does not allow the state to intervene into the allocation of resources or into price setting. Its only intervention involves running institutions allowing market transactions which would otherwise not be possible. It can, for example, provide a legal framework.

At the other extreme, the state has a maximalist role, meaning that it intervenes relatively deeply into a sector economy. It regulates the market forces in order to achieve certain politically established objectives which may relate to prices, allocation of resources and production, entry into the business, etc. The objectives pursued may, for instance, be to ensure that the market participants earn a certain income, that production reaches a certain level, etc. Western states' involvement in agriculture comes close to the maximalist role.

Procedures to approach policy problems define what should be conceived of as policy problems, which of them are most important and how they should be examined (see Jenkins-Smith and Sabatier, 1993, p. 41; Campbell et al., 1989, p. 86; Olsen, 1991, p. 93). Such procedures are used by network members to handle the ongoing flow of policy problems. The procedures affect the nature of the problems put on to the network's agenda; those which the established procedures do not prescribe ways to resolve are excluded, while those which can be handled by the use of established procedures are included. Furthermore, the procedures prescribe how problems accepted on to the agenda should be examined, what counts as evidence and which analytical methods are acceptable. Consequently, they also define which resources actors view as valuable in the policy process. Policy making in economic sectors, for example, values systematic information on economic or technical variables; ideological arguments are, to a large extent, regarded as irrelevant to the policy process.

Marsh and Rhodes (1992; Rhodes and Marsh 1992a) have developed a continuum upon which various policy networks can be placed. This book maintains this notion of a network continuum but uses a set of dimensions which differs from Marsh and Rhodes'. In figure 2.1, policy communities and issue networks are the two extreme network types on a continuum.

Table 2.1
Extremes on the policy network continuum

Dimensions	Policy community	Issue network
Membership	Very limited number of members. Narrow range of interests represented	Large number of members. Wide range of interests represented
Integration	Bargaining and negotiation. Frequent interaction	Consultation. Unstable pattern of interaction
Institutionalisation	Consensus on policy principles and procedures to approach policy problems	Conflict over policy principles and procedures to approach policy problems

A policy network often has a core and a periphery. In such cases, the core consists of actors continuously involved in the policy process on all matters relating to the policy issue. Members of the periphery are only consulted on specific issues in which they have particular resources (Smith, 1992, pp. 314-5; 1993, p. 61; Rhodes and Marsh, 1992, pp. 191-3). We can also identify policy networks which have a high degree of institutionalisation and integration, although they consist of many members who cover a relatively wide range of interests. For instance, the Danish policy network dealing with the European Union's single market project consists of representatives from industry, the crafts, trade unions, employers and consumers. Although many actors participate, there is a relatively high degree of institutionalisation in the network. Furthermore, bargaining characterises its mode of interaction (Daugbjerg, 1994).

Policy networks and the concept of power

Policy networks cannot be separated from the question of power. They have a major impact on power relations which, in turn, influence the choice of policy (Marsh and Locksley, 1983, pp. 43-4). In the formative phases of a public policy, the dominant actors can shape policy principles in a way which serves their long and short term interests. To apply the concept of power in association with network analysis, it is necessary to distinguish between different types of power.

Classical pluralists operate with one dimension of power (the first face of power). Dahl's definition represents the most commonly expressed view on power within classical pluralism. He describes it in the following way: 'A has power over B to the extent that he can get B to do something that B would otherwise not do' (Dahl, 1969, p. 80).[4] The definition is only concerned with the overt exercise of power. However, power involves more than one dimension (Marsh and Locksley, 1983, p. 22). In their critique of this early pluralist view, Bachrach and Baratz (1962, p. 948) add one more dimension to the concept. Accepting that Dahl's definition covers important aspects of political life, they argue that 'power is also exercised when A devotes his energies to creating or reinforcing social and political values and institutional practices that limit the scope of political process to public consideration of only those issues which are comparatively innocuous to A.' Thus, when A can prevent B from being able to put certain issues on to the political agenda, A also has power. For Bachrach and Baratz this dimension of power is the second face of power. Lukes goes even further and argues that there is also a third face of power. He criticises Bachrach and Baratz for dealing only with types of power which involve actual political conflict. For Lukes (1974, p. 24), power also involves:

... exercise of power to prevent people, to whatever the degree, from having grievances by shaping their perceptions, cognitions and preferences in such a way that they accept their role in the existing order of things, either because they can see or imagine no alternative to it, or because they see it as natural and unchangeable, or because they value it as divinely ordained and beneficial.

In this view of power, people have real interests which they do not express or are not even aware of. It is, however, doubtful whether Lukes really adds much to the second face of power. The use of third face power requires an exerciser of power to manipulate the interests of others. To me, it is what A does in Bachrach and Baratz's second face of power when s/he creates or reinforces social and political values. Compared to the second face of power, the exercise of the third face of power is more a matter of degree than of type.

Martin Smith (1990) develops a model of structural power which overcomes the problems of distinguishing between the second and the third face of power. He differentiates between structural and non-structural power. The former includes aspects of the second and the third face. For Smith (1990, p. 39), '[t]he basis of structural power is that rules, procedures and beliefs support the interests of the powerful without the powerful having to decide on every occasion what should be allowed on the agenda.' These rules, procedures and beliefs give certain interests a privileged position in policy making because '[p]ower is not only exercised by individual decisions but by privileges created by structure which advantage certain groups to the exclusion of others' (ibid., p. 35). Non-structural power refers to the exercise of power arising not from structure, rules, procedures and beliefs, but from situations in which actors apply their resources and skills. This type of power is similar to the first face of power (ibid.).

The concept of structural and non-structural power is closely related to policy networks. Policy networks distribute power unequally among political actors. Actors who are members of networks are more privileged than those who are excluded. In policy communities, actors rely on the exercise of structural power in that certain groups do not have access to the policy process. Power is also structural in the sense that within networks, actors' choices are constrained since certain options are excluded whereas others are favoured. Thus, policy communities create power relations which privilege some groups over others. In contrast, the exercise of non-structural power characterises policy making in issue networks to a much greater extent than in policy communities (Smith, 1993, p. 73). Once outsiders place the question of new, unwelcome policies and policy reforms on the political agenda, their opportunities for bringing about change depend upon the type of power that members within existing networks can mobilise to oppose changes. Members of policy communities are in a privileged position because they can mobilise structural power, whereas members of issue networks can rely only on non-structural power.

Cohesion in policy networks and the opportunities of outsiders

In this chapter, it is investigated how existing meso-level structures influence the opportunities of outsiders when they try to bring about policy change within an old policy sector. The structural network approach, which is applied here, undertakes meso-level analysis which focuses on the relationship between a government department and interest groups (Rhodes and Marsh, 1992, pp. 1, 12) whether this occurs at the sectoral or at the sub-sectoral level. Meso-level analysis is concerned with the interaction between organisations; however, in practice, individuals represent organisations in the policy process. These individuals are assumed to pursue organisational goals in their interaction with representatives of other organisations. Meso-level analysis examines relationships which are structural rather than personal (Marsh, 1995, p. 2) and it 'concentrates on questions concerning the structure of the network and the patterns of interactions within them' (Marsh and Smith, 1996, p. 13).

Policy networks structure the decision making process and provide outsiders and insiders with different opportunities for changing or maintaining the existing order in a sector (Smith, 1993, p. 97). Outsiders' chances of bring about change 'depends on the *opportunities* embodied in the network' (Döhler, 1991, p. 239). Some configurations of policy networks facilitate changes whereas others prevent them (Lembruch, 1991, p. 125). The key to revealing the different opportunities for policy change embodied in a network is to analyse the degree of cohesion within it. Cohesion refers to 'the sharing of meaning attached to the interactions within the network' (ibid., p. 126). For those wishing to maintain the existing order in a policy sector, a cohesive meso-level policy network is a very powerful political resource. On the other hand, it is a major problem for outsiders. They may be forced to pay a high price for a compromise or they may even fail completely; however, policy learning may strengthen their position through time (Jenkins-Smith and Sabatier, 1993).

Cohesion among members of a policy network is associated with the network's character. While cohesion characterises policy communities (Stones, 1992, p. 201; Grant, 1992, p. 58), it is absent in issue networks and other types of loose networks. The development of cohesion is facilitated if the membership of the network is restricted and if the degree of integration is high. The fewer members, the fewer interests have to be reflected by the policy principles and thus the more manageable and predictable the policy process becomes (Daugbjerg, 1994, p. 462). Hence, the members can more easily achieve consensus among themselves. Cohesion does not develop when the network members' objectives are in conflict; that is, when 'the realisation of one actor's objective impedes the realisation of the other actor's objective' (Bressers, 1995, p. 4). Usually, networks with many members are characterised by actors pursuing conflicting objectives. 'Frequent [and high-quality] interaction between all members of the community on all matters related to the policy issue' (Rhodes and Marsh, 1992a, p. 186) also eases the development of cohesion. Intensive information exchange means that each member achieves insight into the others' situations, interests and values. This makes it possible to achieve a consensus on which principles and procedures should prevail in policy making.

Institutionalisation influences the content of public policies. In the formative phases, a dominant coalition of political actors become core members and use this position to ensure their power in future policy making through the structure of the network. They shape policy principles and procedures to approach policy problems in ways which are advantageous to them. Typically, they lay down policy principles which pass on the costs of policy to groups not represented in the network. Procedures to approach policy problems constrain the policy agenda and shape the policy process in ways which put core members in a privileged position.

Over time, the institutional factors in a policy community become more important in explaining policy choices than the pattern of resource exchange because members increasingly handle policy problems by referring to principles and procedures. The procedures to approach policy problems, which actors lay down in the formative phase of a network, to a large extent define which resources are important in policy making. In other words, the pattern of resource interdependency becomes institutionalised. Actors developing resources which are not defined as relevant to policy making cannot, therefore, gain access to the core of the network. The high degree of institutionalisation in a policy community means that its members' policy preferences, to a large extent, is shaped, or even determined, by its rules, principles and standard operating procedures. It can occasionally be seen that political actors follow institutional principles and procedures even 'when it is not obviously in their narrow self-interest' (March and Olsen, 1989, p. 22).

The roles of policy principles and procedures to approach policy problems become evident once we analyse a policy community's policy decisions over time. The trend in these decisions reflects the structure of the network and therefore policy making remains stable for long periods. In specific decisions, however, principles and procedures of networks do leave room for actors to pursue rationally their objectives. Even the tightest policy communities are not so constraining that their members cannot have different views on policy problems and which specific solutions are attractive. As Windhoff-Heritier (1991, p. 41) points out: 'Institutional structures and rules are rarely ever deterministic, in the sense that they shape behavior fully. Instead they convey general orientations for action ..., leaving room for self-interest and strategic decisions.' In a similar vein, Elinor Ostrom (1991, p. 239) suggests that '[c]hoice from among those actions that are ruled in cannot be made on the basis of institutional rules.' As already stated, making compromises is a dominant feature of the policy process, even in a policy community. Such a process involves tactical manoeuvring in which actors aim at achieving solutions which are optimal for them, but these agree with the network's policy principles.

The high degree of cohesion in a policy community influences its members' formation of policy preferences. In particular, the network structure influences the policy preferences of the state actor most centrally positioned in a policy community. State actors - for example ministries of agriculture - do not have policy preferences as obvious as those of interest groups. Indeed, we cannot take for granted that they have fixed policy preferences.[5] Their formation of preferences is influenced by the context within which they operate. Olsen (1983, p. 145) argues that the views of top civil servants 'are formed by the tasks they are responsible for, by the institutions and the professions to which they belong, and by the parts of the environment they interact with.' In meso-level policy making, state actors' 'interests develop from contacts with groups, and often within networks' (Smith, 1993, p. 227). The interests of the state actor and a particular organised interest may thus converge. Groups' interests are, to a greater extent, derived from their position in the economy or from the functions they perform, but also from perceived organisational needs and from the network structure.

When demands for policy change in a policy sector appear on the political agenda and can no longer be avoided or downgraded, members of a policy community are forced to act. Because it comes naturally to them, they will apply well-known methods and policy principles to solve the new problems. The experience derived from earlier decision making processes, thus, forms the basis of their approach (March and Olsen, 1989, pp. 22, 38, 167-8; Rhodes, 1981, p. 118; see also Hall, 1986, p. 233). Members of a policy community eventually match a situation to the demands of established policy principles (March and Olsen, 1989, p. 23). A policy principle survives because actors perceive it as an obvious way to cope with certain problems although these have a different character than those which originally gave rise to that principle (Rothstein, 1988, p. 38; see also Krasner, 1984, p. 240).

Outsiders' opportunities for forcing new policy principles into a sector in which a policy community exists are limited. They will meet strong united resistance. The state actor which is a member of the established network allies with the interest group which is its traditional negotiating partner. The cohesion is in some cases so strong that the network can be said to act in unity (Daugbjerg, 1994a, p. 13). Cohesive policy networks, thus, favour the interests of the group subject to policy change since it can rely on structural power to oppose outsiders; that is, its power is derived from the structure of the network itself. It can mobilise strong opposition against outsiders because the consensus on principles and procedures facilitate coalition building within the network. In such situations, outsiders have limited opportunities to ensure that their policy principles become the basis of policy change. Therefore, established principles are most likely to continue to underpin policy.

Members of a cohesive policy network may, however, give some concessions to outsiders to avoid a situation in which the policy process develops into a highly politicised conflict. Only rarely can members of a cohesive network totally dismiss demands for policy change because there is a risk that the issue will stay on the agenda for a long time with the effect that new actors supporting policy change are mobilised. This threatens the policy network's control over the policy process which may, if the control is lost, move from the bureaucratic arena to the cabinet and perhaps to parliament. Policies adopted in these two arenas may - due to a high level of public attention - result in outcomes which the network's members cannot accept (see Schattschneider, 1960, pp. 15, 36). Therefore, network members have a great incentive to give some concessions to ensure their control over the policy process. The concessions are not, however, likely to introduce new policy principles into the sector concerned and, consequently, policy changes will be moderate. It must be said, however, that although network members suffer a political defeat in the policy formulation process, the battle is not lost. As public attention decreases, they may regain their power, enabling them to enter the political-administrative process and turn the implementation process in a direction which limits the perceived disadvantages of the policy.

Members of loose networks react differently to demands for changes. Issue networks are characterised by low degrees of institutionalisation. There is no dominant coalition possessing the power to lay down a set of policy principles and procedures to approach policy problems which can guide policy making. Members of an issue network constantly disagree over principles and procedures. The low degree of institutionalisation implies that policy making is primarily guided by resource exchange from issue to issue. Members occupy central positions in issues in which they have resources needed by other members whereas they are of secondary importance in issues in which they do not possess such resources. The lacking consensus on principles and procedures makes it difficult for issue network members to make policies which are more than the 'lowest common denominator.' There is among them no basic agreement on which types of solutions are attractive and on the way to reach agreements (Smith, 1993, p. 64). Actors enter the policy process with a variety of policy solutions. Controversies within the network evolve along several dimensions, making it very difficult, if not impossible, for the members to define which policy principles should be the most important. Further, members have different views on what should be defined as problems. Which resources are relevant is also open to dispute. Therefore, the policy process in an issue network is characterised by actors who try to convince each other that they have the most important resources. Issue networks are distinguished from the general interest group universe by a minimum degree of exclusion (ibid. 65). Although access to an issue network is relatively open, there are some restrictions. One of these are that actors must have 'legitimate' interests in the issues addressed. In addition, relationships are, to a limited extent, ordered (Rhodes and Marsh, 1992a, p. 190).

In an issue network, the pattern of resource interdependency plays the major role in policy making. Because there is conflict over the principles and procedures, the resources available to each participant in the network are important factors in policy making (Windhoff-Héritier, 1991, p. 42). There are only a few limitations to the use of resources; actors can use them in the way which seems most advantageous. If actors think they can further their interest by using mass media or applying high profile campaigns, they will do so, even though such moves hurt the relationship between network members and, thus, have a negative impact on future policy making.

Due to the constant conflict over principles and procedures, no cohesion among the members will develop. To decrease uncertainty network members may make compromises on some principles, but the network is, nevertheless, basically unstable because it is not based on a shared understanding (see Smith 1993, pp. 126-7). However, these compromises may bring about a limited degree of order within the network.

When members of loose networks confront demands for change, they continue to be in disagreement over which policy principles to apply in policy making. Each member, or group of members, pursues what it perceives as its self-interest, trying to achieve support for policy principles which would benefit them in present and future policy making. Members of loose networks do not, therefore, possess structural power; they have to rely on non-structural power. Since access to the policy process is relatively easy and since there is no consensus on which policy options are attractive and feasible, members use their resources and skills to influence policy outcomes.

A low degree of cohesion means that outsiders have better opportunities to bring about policy changes in a sector in which they do not traditionally participate in policy making. Not only can they ensure that problems are being dealt with, they also have good opportunities to gain sufficient political support for their policy principles. In policy networks with low degrees of cohesion, members have difficulties in forming strong coalitions. Usually, they can only form a weak alliance of resistance, if any. The interest group whose members will bear the costs of policy change is, therefore, left without centrally placed political actors (particularly state actors) which can support it in the policy process. State actors within established networks either become mediators between outsiders and the group subject to policy change, they pursue their own bureaucratic interests or they do both. Because a strong status quo minded coalition cannot be formed, policy change tends to be based on policy principles which are significantly different from those already established.

The cohesion of networks also influences the strategies of political actors. Cohesive networks enable their members to form a strong coalition defending the status

50

quo and thus to pursue a confrontational strategy when dealing with outsiders. Outsiders may be satisfied with minor alterations in such situations because they realise that the structure of the network provides only limited opportunities for change. The low degree of cohesion in loose networks rarely enable their members to unite against outsiders. Realising that such networks provide favourable opportunities for change, outsiders may choose a strategy which aims at fundamental policy change. Those who are subject to change are in a weak position. They pursue, therefore, a cooperative strategy in order to ensure that they have some influence on policy choices. Making concessions to outsiders is preferable to exclusion.

Conclusion

In this chapter, it has been argued that many network analysts confuse structural features of networks with properties of the agents and with characteristics of policy outcomes. This impedes the development of a network model explaining policy choices. To overcome that problem, this chapter has developed a policy network continuum which applies membership, integration and institutionalisation as the analytical dimensions. Membership refers to the number of participants and the range of interests they represent. In integration, the continuum distinguishes between the two extremes bargaining/negotiation and consultation. Institutionalisation refers to whether there is a consensus on policy principles and procedures to approach policy problems. A policy community is a highly integrated and institutionalised network in which membership is very restricted. By contrast, in an issue network, the policy process is relatively open and, as a result, many actors participate. The degrees of integration and institutionalisation are low.

The network continuum is the basis for the development of a network model which can help to explain policy choices. The model developed here links policy networks and the choice of policy in situations where an actor outside a network has forced demands for change on to the networks' agenda. Cohesive policy networks provide an unfavourable opportunity structure for outsiders because members of such networks are bound together by a shared understanding of which policy principles are attractive. When outsiders threaten the established order, members of a cohesive policy network can, thus, build a strong coalition opposing policy change. The strength of the coalition depends on the policy preferences of the state actor most centrally positioned within the network. Since state actors' interests are not fixed, network structures influence their formation of preferences. In cohesive networks, state actors' policy preferences come quite close to those of interest groups. For interest groups a cohesive network produces structural power because it enables them to build a coalition with state actors which can limit outsiders' opportunities for bringing about fundamental policy change. Since demands for policy change can rarely be totally dismissed, policy makers are likely to bring about moderate policy changes which do not involve a shift in policy principles.

Outsiders' situation is better if a non-cohesive network exists in a policy sector. Such networks provide opportunities for change because their members are not

51

bound together by a consensus on policy principles. Policy making is guided by the patterns or resource interdependence from issue to issue or even from question to question. Coalitions against change are likely to be weak. When faced with pressure from outsiders, the state actor most centrally positioned in the network often chooses to pursue its own bureaucratic interests and/or becomes a mediator between outsiders and those who are the targets of policy change. Therefore, interest groups opposing change cannot use the network itself as a power base; they have to rely on non-structural power. Policy changes tend to be fundamental because new principles are introduced to underpin the choice of policy. Quoting Peter Hall (1986, p. 266), one can conclude that a certain 'organizational structure tends to lead policy-makers into some courses of action and away from others; and each course of action tends to favor the interests of some social groups over others.'

Political actors do not only derive their power from meso-level structures. Certain features of the political systems to which they belong also work in favour of some interests and against others. The following chapter deals with the role of macro-political variables.

Notes

1. However, an actor's resources must be seen in relation to those of others. A's resources may be the structural constraint of B.

2. The dimension can also be an outcome if the network affects actors' choice of strategies by precluding some strategies and making others possible.

3. For examples, see Jordan and Schubert (1992), Atkinson and Coleman (1989) and van Waarden (1992).

4. First published in 1957.

5. This is precisely why rational choice institutionalism has limitations in explaining the policy preferences of state actors in public policy making. As Tsebelis (1990, pp. 32-33) points out: 'As the actors' goals become fuzzy ... rational choice explanations will become less applicable' (see also Hall and Taylor, 1996, p. 951).

3 Moving beyond meso-level analysis

Introduction: how to move beyond meso-level analysis

In the preceding chapter, I examined the role of meso-level policy networks in cases in which outsiders force change on the agenda. In other words, I analysed meso-level structures. This analysis, however, only provides partial explanation. As Atkinson and Coleman (1989, p. 67) suggest: '[m]eso-level phenomena cannot be explained in isolation from broader political institutions' (see also Marsh, 1995; Lembruch, 1991, pp. 134-5). These broader political institutions affect the opportunities for policy change.

There are two approaches for integrating policy network analysis with macro-level analysis. One begins from macro-theories and applies meso-level analysis as a method of supporting macro-theories. The other puts network analysis at the centre and attempts to identify the features of the broader context which have an impact on meso-level processes. State theory is the point of departure of the former approach. Marsh (1995) and Marsh and Daugbjerg (1998) argue that state theory can inform us on the overall relationship between the distribution of power in a society and the formation of meso-level policy networks. Revealing this link and thus characterising whole political systems is a major strength of such an approach. Clearly, the weakness of the approach is that it has difficulties explaining inter-sectoral variation and sectoral variation among countries in which the overall distribution of power is similar (Marsh and Daugbjerg, 1998). Since the purpose of this book is to analyse cross-national variation in policy outcomes, the approach beginning from macro-level theories cannot provide a sufficient explanation.

The alternative is to put meso-level analysis at the centre and search for macro-level variables which have a direct impact on the meso-level processes. Rhodes (1992) uses such an approach in his study of sub-central governments in Britain. He applies macro-level analysis for exploring 'those features of the national government environment which directly impact on the sub-central system...' (p. 48). 'National government environment' refers to 'central government institutions and their socio-economic environment as they impact on [sub-central government]' (ibid.). He uses several characteristics to describe the national government environment in Britain. These features are:

> an unstable external support system; the decline of the mixed economy; the growth of the welfare state; the extension of the allied process of functional differentiation and professionalization; the development of a social structure characterized by multiple (non-class) cleavages; and the continuing stability of a political tradition characterized by a two party system, a unitary institutional structure and a central élite ideology defending the mixed economy welfare state (ibid., p. 49).

The strength of Rhodes' work is that it outlines a method for the integration of meso-level and macro-level analysis. Its weakness is that it is under-theorised. Since there is no shortage of macro-variables which influence meso-level processes (Atkinson and Coleman, 1992, p. 166), a macro-model needs to include only those variables which have most influence. Additionally, the model must accurately define the variables. More importantly, it must also be capable of suggesting how these variables influence meso-level policy making, in particular of specifying the conditions under which a particular outcome is likely to occur or not (ibid., p. 167). In other words, we need to establish the causal links between the variance in the macro-variables and the variance in the meso-level variables. It would probably be too optimistic to expect anyone to develop a theoretical model which specifies how all possible values in the macro-variable are linked to certain outcomes in the meso-level variable. Hence, initially, we should be satisfied with propositions suggesting the outcomes which the extreme values in the macro-level variable produce at the meso-level. But further research must lead us towards more precise theoretical models.

There has been a tendency in network analysis to examine the policy process without including the broader state institutions in the study (Atkinson and Coleman, 1992, pp. 155, 168). If one's research interest is limited to identifying and describing policy networks, there is no need to integrate these macro-political institutions into the analysis. Furthermore, if one adheres to the traditional pluralist view of the state as a highly fragmented and open set of bureaucracies competing with each other, then the state is no more than 'a complex entity spanning multiple policy domains, comprising both government organizations and those core private participants whose interest must be taken into account' (Laumann and Knoke, 1987, p. 381). The distinction between state and society thus disappears (Atkinson and Coleman, 1992, p. 164), implying that 'the appropriate unit of analysis for studies of policy formation is not the state understood in institutional terms, but the state as a collection of policy arenas incorporating both governmental and private actors' (Laumann and Knoke, 1987, p. 7). This pluralist perception of the state does not require the researcher to put much emphasis on the way in which macro-political variables influence meso-level processes; however, by removing the influence of macro-political variables from the analysis, major explanatory variables are excluded. For instance,

Smith (1993) argues that different state structures help to explain why British agricultural policy making has remained within a tight policy community whereas in the United States, it occurs within a policy network sharing several features of an issue network. Thus, if the researcher wants to explain policy outcomes, particularly variation in outcomes across sectors or countries, s/he can hardly neglect the influence of broader state institutions. Therefore, network analysis must leave 'some room ... for the reintegration of macropolitical structures into the analysis of policy outcomes' (Atkinson and Coleman, 1992, p. 163). These macro-political structures constrain or facilitate actions at lower levels (see Ostrom, 1990, pp. 51-2).

In this chapter, I identify two macro-political features which characterise a political system as a whole and which, at the same time, affect meso-level policy making. I shall argue that a social group's structural power in parliament[1] and the organisational structure of the state influence meso-level policy processes. They have an impact on the creation of meso-level policy networks and on the choice of sectoral or sub-sectoral policies. Only few studies analyse the theoretical link between political parties and the choice of policy. Lembruch (1979), for instance, is one of the few corporatists who shows that political parties do, in fact, play a role in public policy making. He argues that parties are best suited to producing long term policy decisions (p. 156). Steen (1985; 1988, ch. 8) also deals with the role of political parties in public policy making. He analyses how political parties influence the contents of agricultural policies in Norway, Sweden and Great Britain. Unfortunately, his analysis is underdeveloped theoretically.

The theoretical model developed in this chapter concerns policy making within nation states. Since one of the empirical cases being examined is a reform process in the European Community, it will be shown how we can apply the model in studies of policy making at the EC level.

Do parties and parliaments make a difference in public policy making?

Pluralists, corporatists and network analysts have not paid much attention to the roles of parliaments and political parties in public policy making. They all tend to see the policy process as non-parliamentary (Judge, 1990, ch. 2-3; 1993, ch. 4).[2] Grant Jordan (1990, p. 473) is one of the scholars who argues that parties and parliaments have insignificant roles in policy making. As he points out: 'The notion that policies are the product of sectoral bargaining in which ministries have clientelistic orientations to the major groups is the conclusion of a trend in political science that has also de-emphasized the importance of legislatures, parties and the formal institutions of government.' Writing about Britain, Jordan and Richardson (1987a, p. 57) try to make this point even more clear by suggesting: 'whether the House [of Commons] contributes more to the policy process or to the tourist trade is a difficult question.' The limited role of parliaments is also expressed in the 'decline of legislature thesis.'[3] In public policy making, the thesis suggests that:

The specific measures of public policy are formulated by government departments following consultations with affected interests. Those measures are presented to the legislature for approval. The legislature will usually lack the political will to challenge them: a partisan majority will exist to pass them. It may also lack the resources to challenge them: if presented with a measure that has the support of the different groups affected by it, it may have no alternative source of information or advice to challenge the agreed package. Consequently, the measures are passed (Norton, 1993, pp. 3-4).

Though parliaments may not have the resources to change specific measures in a public policy, they certainly have the power to disapprove them and force governments to come up with other measures. Indeed, no minister or government wants to face defeat in parliament (see Brand, 1992, pp. 82-6). Rejecting the notion of parliaments as powerless bodies, Judge (1993, p. 124) points out: 'To conclude that the legislature's substantive contribution to law making is limited, even peripheral in the case of detailed formulation and implementation, does not mean that parliament is peripheral to the process of policy making itself.'

Parliament's views are considered in advance and, thus, its role in the policy process is indirect in most respects. Writing on the Norwegian parliament (the *Storting*), Olsen et al. (1982, p. 65) suggest: 'The influence of the Storting is often manifested not by active participation in the policy process, but by how its interests and views are taken into account in advance by other participants in the policy process' (see also Christensen, 1980, p. 360; Judge, 1990, p. 58; Olsen, 1983, p. 39). So, the 'core defining role [of parliaments] is not to *make* law but rather to *give assent* to it' (Norton, 1993, p. 5; see also Brand, 1992, p. 73). Overlooking the role of parliaments in analyses of public policy making is, therefore, a mistake (Brand, 1992; Judge, 1990; 1993). Indeed, parliaments have a significant influence on public policy making. What goes on in a meso-level policy network 'is *structurally constrained* by the parliamentary system' (Judge, 1993, p. 125). The notion that legislatures have very limited influence is based on conclusions arising from the application of inadequate research methods. Such methods focus on the observable behaviour of parliament; thus they only reveal the exercise of the non-structural power (the first face of power). The power of parliaments also rests on the exercise of structural power (the second and third faces of power).

Since representative democracy is the major character of governance in Western societies, it does not make sense simply to argue that parliaments and political parties are excluded from influence because we cannot observe their direct effect on the policy outcome. The influence of parliament arises from the actions of the parliamentary parties because '[p]arty remains the basic determinant of parliamentary behaviour' (Norton, 1991, p. 80). Relatively few political scientists have studied the relationship between interest groups and parliamentary parties (Grant, 1993, p. 125). Insofar as such studies have been undertaken, some focus on the way in which interest groups can use political parties to gain influence and on the limitations of using the party channel of influence (e.g. Jordan and Richardson, 1987, ch. 10; Grant,

1993, ch. 7; Wilson, 1990, p. 6). Others attempt to map the interactions between parties and organised interests and to reveal their importance (e.g. Damgaard, 1982, 1982a, 1986; Norton, 1991). In the latter type of studies, there seems to be an implicit assumption that the pattern of interaction reflects the influence of various interest groups. It may be so; however, their influence may not only be reflected in the pattern of interaction. No doubt the exercise of non-structural power is an important aspect of interest groups' attempts to influence political parties but the structural power arising from the political parties' sympathy towards various social groups is an even more important factor explaining interest groups' influence in the policy process. Therefore, in revealing how the political parties privilege certain interests, and not others, we can examine their ideologies, general policy positions and electoral appeals. These bias the policy process in a direction which favours some groups over others.

The theoretical model developed here can only be applied in parliamentary democracies[4] with cohesive political parties which have high degrees of internal discipline. Before I proceed, it is important to point out that political parties are not merely agents of special interests. Even the most narrowly focused parties must deal with broader issues, particularly if they are in power (see Jordan and Richardson, 1987, pp. 243-4). As Wilson (1990, p. 156) points out: 'No political party can win a majority by being identified with a single interest' (see also Grant, 1993, p. 126). As a result of their claims to democratic legitimacy, political parties generally enjoy some autonomy in their relations with interest groups.

The degree of sympathy which political parties give to the interests of a social group is an important determinant of the structural power of an interest group. For instance, farmers have benefited from the privileged position they have enjoyed within Labour and Social Democratic parties as well as in agrarian parties or those which traditionally attract farmer votes (Wilson, 1990, p. 165; see also ch. 8 in this book). Strategies for persuading politicians to support them play a limited role in explaining why farmers have achieved political power that is disproportional to their number of votes. Social groups may enjoy either wide support among political parties or support only from one or a few parties. A social group in the latter situation is not necessarily in a weak position. A party which protects that group's interests may have a strategic position within the party system, enabling the party to deliver the decisions which the group demands.

Parties influence the creation of meso-level policy networks (Damgaard, 1986, p. 275). The structure of party loyalties has an impact on who is excluded and who achieves insider status. If an interest group has wide or strategically positioned support within parliament and an opposing interest group does not, then it is unlikely that both groups will achieve equally strong positions within a meso-level network. The most likely outcome is that the interest group favoured by parliamentary support will form a policy community with the state actor responsible for the policy area concerned. In contrast, a policy community will not be formed if a group's interests are not sufficient protected by the political parties. Nor will a policy community be formed if an opposing interest group has counterbalancing power in parliament.

Instead, a policy network in which both conflicting interests are represented will be constructed. If there are many conflicting interests, an issue network will emerge. Parliamentary support does not, however, need to be as direct as suggested here. Occasionally, bureaucrats and interest group representatives can form policy networks without the involvement of parliament. This is possible when parliament does not oppose the network construction.

Although one should not overlook the role of parliament in the formation of meso-level policy networks, it is important not to overestimate it. Parliamentary support is a necessary but not a sufficient factor determining their nature. Parliament may give an interest group access to a network but this does not necessarily mean that the group gains a central position because it may not possess the resources which are valued by the other network members. What parliaments can do is to give interest groups access to a policy network. However, the pattern of resource interdependencies determines, to a large extent, their position within the network.

In some circumstances, party change may be an important source of network change. If the network produces outcomes which are acceptable to groups and state actors outside the network, or at least do not bring about opposition, party change may not have any immediate consequences for it. But if the outcome becomes a source of opposition, party change may have an impact. If the groups favoured by the outcome produced by the network have lost parliamentary support because of party change, pressure for reform can lead to a reorganisation of the network so that new interests are included.

Policy change and structural power in parliament

Attempts to bring about policy chance in old policy sectors cannot be successful if a majority in parliament cannot be formed. Majorities can only be created through inter-party, inter-factional or inter-personal coalition building. This is often difficult because the stability of old policies is based on the structural parliamentary power of one or very few interest groups. However, over time, this structural power may decline. A social group's structural power in parliament refers to the extent to which political parties favour that group's interests in their general policy positions and electoral appeals.

Only a historical analysis can reveal whether the structural power of a social group has declined. At the time when politicians approve the policy principles of a public policy favoured by certain interest groups, the structural parliamentary power of these groups is strong. In some cases, they may remain powerful for decades; in other cases their power decreases. A decline need not lead to immediate change in policy principles. Once policy makers have made a historical choice on the principles and have formed a tight policy community, future changes become difficult because such a network protects the established principles. However, even the most tight policy communities cannot exist independently of parliament. In a parliamentary democracy, the political executive is required to answer to parliament (Judge, 1993, p. 125). Therefore, decline in a social group's structural power in parliament can lead to policy change.

Pressure for change may emerge if some organised groups or state actors who do not belong to the network's core believe that the established policy causes too many political and economic disadvantages (see Brand 1992, p. 78). A policy may not produce costs immediately after it has been put into effect but, as time goes by, it may begin to produce anomalies which mobilise some groups and state actors (Smith, 1993, pp. 92-3). Whether these actors can set the agenda for policy change and subsequently carry it through, depends on how much the structural parliamentary power of the social group benefiting politically and economically from the established policy has declined. If the erosion is significant, there is an opportunity to reform old policies and to introduce new ones.

A social group does not derive its structural power in parliament from its electoral power alone. Structural power often has historical roots. The formative phase in a party's history continues to condition its political behaviour in many ways (Panebianco, 1988, pp. xiii, 63). A party's 'original official aims are never abandoned, nor do they ever become a mere 'facade' ... [It] continually engages in certain activities related to these aims, for it is precisely upon these activities that the party's collective identity and the leadership's legitimacy are based' (ibid, p. 16). However, parties may change since they are vulnerable to changes in social structure. The concomitant of an increase in voters in one social group is a decrease of voters in another. The decline in the number of core voters, due to social changes, presents a party with a dilemma. It has to choose whether to remain true to the interests of its traditional voters and face declining electoral support or to transform and attract votes from new social groups in order to maintain, or even increase, the electoral support. Parties react differently to changes in social structure. Some transform whereas others remain loyal to their core voters (Lane and Ersson, 1991, p. 112). Although the former, as catch-all parties, try to attract new voters, they still hold on to their traditional supporters (Kirchheimer, 1966, pp. 185-6, 193).

The way political parties deal with social changes influences their responses to new agendas within a policy sector. Some become a source of change because 'contending political parties have an incentive to search for new solutions to old problems. The ability to offer such solutions is an important source of electoral appeal' (Hall, 1986, p. 273). Other political parties oppose changes in a policy sector because alterations will go against the interests of their core voters. Another source of conservatism is party ideology. As Peter Hall (1986, p. 272) points out: 'Like all organizations, [parties] become attached to a particular ideology because ideology is central to organizational life. Ideology provides a political organization with an identity over time, tying present activities to past ideals.' Therefore, a party will not suggest changes in a policy sector just because it finds it opportune so to do. Suggesting or supporting changes which conflict with a party's ideology may produce a crisis within the party, resulting in internal conflicts. Adhering to policy positions which are in accordance with the established party ideology produces coherence within a party - a highly valued goal of party organisations. Whether or not parties choose to adhere to traditional policy positions has an important impact upon an interest group's structural power in parliament.

The model developed here focuses on multi-party systems, including those with and without a dominant party (Blondel, 1968, pp. 183-9). To establish a social group's structural parliamentary power in a multi-party system, we need to examine the extent to which parties in parliament support the interests of the social group. In addition, we should examine the parliamentary position of the supporting parties.[5] Such an analysis involves four steps, revealing:

- whether parties which are loyal to the social group concerned are present;
- such parties' degree of loyalty to the social group concerned;
- the strategic parliamentary position of such parties;
- the strength of the social group's interests within the dominant party, if such a party exists.[6]

A party's general loyalty to a social group refers to the extent to which it is willing to sacrifice the interests of that group in order to obtain electoral support from other groups. We can assess the strategic position of parties in parliament on their coalition potential which refers to 'the extent to which a party may be needed as a coalition partner for one or more of the possible [and feasible] governmental majorities' (Sartori, 1976, p. 122). The parliamentary position of a party is, therefore, much more influenced by its coalition potential than by its number of seats in parliament. As Sartori (1976, p. 122) puts it: '[a] party may be small but have a strong coalition-bargaining potential. Conversely, a party may be strong and yet lack coalition-bargaining potential.' The strength of a social group's interests within the dominant party can be assessed by analysing both the extent to which that party appeals to the social group concerned, and the extent to which that party perceives the interests of the social group concerned as coinciding with the party's ideology and its core voters' broader interests. The structural parliamentary power of a social group is *highest*:

- when the social group's supporting party is generally loyal to the group's interests;
- when the supporting party has a high degree of coalition potential;
- when the dominant party is sympathetic to the social group's interests.

Conversely, a social group's structural power in parliament is *lowest*:

- when the social group's supporting party tries to attract votes from other social groups by sacrificing the interests of the social group concerned;
- when the supporting party has a low degree of coalition potential;
- when the dominant party is not sympathetic to the social group's interests.

It has already been established that in processes of policy change, actors struggle over the policy principles with which to underpin the choice of policy. In situations

60

in which those subject to regulation view existing policy principles in a positive light, they defend them. The most radical policy changes are those in which outsiders force new policy principles into a sector, whereas established principles remain the basis of policy choices in moderate policy changes. The model developed here addresses situations in which the group facing pressure is happy with the established policy and does not view proposed policy changes kindly.

The composition of parliament influences policy making at the meso-level. An interest group representing a social group with a high degree of structural power in parliament can prevent reformers from introducing fundamental policy change. The parliamentary majority coalition which supports the interest group must be broken if new policy principles are to be introduced into an old sector. Attempts to do this fail if interest groups opposing change are able to utilise their parliamentary support skilfully[7] When such interest groups can mobilise sufficient support in parliament, the established policy principles remain the basis of policy change. In other words, moderate policy change is likely to occur.

If a social group's structural power in parliament has declined, fundamental policy change may occur because the old parliamentary coalition originally supporting the principles of the established policy is easy to break up. Some parties in the coalition have no difficulty in shifting their loyalties from the social group opposing change towards those wishing radical change (or the interests they attempt to further). Such parties do not see the interests of their core voters being hurt by the shift in policy position, nor do they have any ideological problems in changing policy position. As a result, principles which are fundamentally different from those of the established sectoral policy are likely to become the basis of policy change.

Since one of the empirical cases which is examined is a reform process in the European Community, it is necessary to analyse whether the macro-level model on structural power in parliament can be applied at EC level. The parliament of the European Community has a very different role compared to that of national parliaments. While national parliaments have legislative power, the European Parliament (EP) is primarily an advisory body. The actual legislative body of the European Community is the Council of Ministers (Lodge, 1989, p. 42).

Studying the formal powers of the EP may lead one to think that it is a strong body, but a closer look reveals that it has only limited power. As Meester and van der Zee (1993, p. 140) point out:

> At first sight, the powers of the European Parliament (EP) look rather impressive: (i) it has extensive budget responsibilities; (ii) it can dismiss the Commission (i.e., the Commissioners) by a two third majority of the votes; (iii) it may pose written and oral questions to the Commission and may call it to account in plenary or in committee debates; (iv) it can pass resolutions; (v) it is called upon to deliver opinions on draft legislation of the Commission under the consultation and, more recently, the cooperation procedure. ... However, the Parliament still lacks powers of compulsion and, as a result, its influence in EC decision-making is, though it has grown over the years, rather limited.

In agricultural policy making, the EP's power is very limited, both formally and informally. Article 43 of the Treaty of Rome specifies its power in the Common Agricultural Policy area. The article states that the Parliament has only a consultative role (De europæiske Fællesskaber 1987, p. 256). Therefore, the EP's views carry little weight because the Commission and the Council can ignore it; however, the Commission does sometimes take the EP's views into consideration (Meester and van der Zee, 1993, p. 140). Most observers agree that the EP has limited influence on the Common Agricultural Policy (e.g. Atkin, 1993, p. 68; Grant, 1997; Fearne, 1991a, p. 106; Nedergaard et al., 1993, pp. 33-4; Nugent, 1994, p. 382; Phillips, 1990, p. 119; AE no. 1486, 10 April 1992, p. E/3). To achieve influence, the EP sometimes drags its feet and puts 'pressure on the Commission to include changes and propositions from the Parliament in its proposal to the Council' (ibid., p. 141). Since the EP clearly is not a crucial macro-level variable contributing to the explanation of reform outcomes in EC agricultural policy making, there is no reason to include it in the empirical macro-level analysis.

State structures

The starting point for the analysis of another crucial aspect of the broader context influencing sectoral or sub-sectoral policy change is the literature on the state. Although this literature focuses on the national level, it is also relevant when one analyses policy making in the European Community. Like the state, the EC has a set of political institutions which relate to each other in a particular way. The way in which they do this influences policy making.

From a structure-agency perspective, there are two ways of viewing the state. While some writers view it as an agent (e.g. Nordlinger, 1981; Thomsen, 1996), others view it as a structure (e.g. Hall and Ikenberry, 1989). Skocpol (1985, pp. 27-8) tries to bridge the gap between the two positions by arguing that the state can be both; however, her contribution does not resolve the confusion. The state-centred approach views the state as an agent. For Nordlinger (1981, p. 11), who advocates a strong version of the state-centred approach (Thomsen, 1996, p. 39), the state refers 'to those individuals who occupy offices that authorize them, and them alone, to make and apply decisions that are binding upon any and all segments of society.' In an attempt to avoid the confusion of the state debate, one can distinguish between state and state actors. Then, the state refers to a 'set of institutions ... [which] are at the centre of a geographically-bounded territory ... [and which] monopolises rule making within [the nation state's] territory' (Hall and Ikenberry, 1989, pp. 1-2). Modifying Nordlinger's (1981) definition of the state, I define a *state actor* as a group of individuals, or an individual, who occupies an office which authorises it, or her/him, to make and apply decisions on behalf of the state which are binding upon any and all segments of society.

Studying the state is important because its 'organisational configurations ... encourage some kinds of ... collective political actions (but not others), and make possible the raising of certain political issues (but not others)' (Skocpol, 1985, p. 21; see also Katzenstein, 1978; Atkinson and Coleman, 1985; Skocpol and Finegold, 1982). An important organisational configuration which influences meso-level policy processes is the internal division of authority within the state. It has an important impact on the power of state actors vis-à-vis societal interests. Katzenstein (1978, p. 323) argues that the more centralised a state, the stronger it is vis-à-vis organised societal interests; and the stronger the state, the more it can intervene in a nation's industrial sectors.[8] Katzenstein's strong-weak state distinction has been criticised for being oversimplified; reality is often much more complex than he suggests. Research shows that it is common to find strong state actors in states which are usually regarded as weak. Skocpol and Finegold (1982), for instance, provide a clear example, demonstrating that strong state actors can exist within weak states. They show that the US Department of Agriculture was a strong actor in an otherwise weak state when the New Deal agricultural policies were made. In their own words, the US Department of Agriculture was 'an island of strength in an ocean of weakness' (p. 271). The opposite phenomenon is just as common (Wilks and Wright, 1987, pp. 281-5; Atkinson and Coleman, 1989, pp. 48-9). Grant, Paterson and Whitston (1988, p. 307) also question the strong and weak state distinction. They find that the 'differences between British and German practice ... in the chemical industry ... are not as great as a reading of the literature on government-industry relations in the two countries might lead one to suppose.' One should expect marked differences in policy outcomes because '[t]here is a substantial literature which suggests that West Germany has greater capability than Britain to deal with problems of industrial adjustment' (ibid.).[9]

In an attempt to overcome the problems with Katzenstein's strong-weak state distinction, Atkinson and Coleman (1989; see also 1985) apply it at the meso-level, but point out that we should not be tempted to lose sight of macro-level analysis (1989, p. 49). By focusing on the meso-level alone, Atkinson and Coleman do, indeed, overlook the fact that the overall structure of the state influences the opportunities for coordination. States in which authority[10] is dispersed are likely to develop competing decision making centres which constrain the power of state actors interacting with interest groups. In such a situation an interest group can play competing state actors off against each other. It can also take advantage of the situation if a competing state actor allies with it in order to achieve control over the policy process. In both cases, the interest group can exercise considerable political influence. The least centralised states tend to be federal states and states with strong parliaments which limit the power of the executive. Such states develop competing decision making centres. Fragmentation emerges from competition between national and regional (or sub-national) decision making centres, from competition among state actors at the same administrative level of the state and from checks-and-balances between the legislature and the political executive. Division of authority occurs as a result of constitutional or informal rules of governance.

State structures influence the formation of meso-level policy networks. State actors which can aggregate authority within the state are powerful and can construct policy networks including two opposing interests (Atkinson and Coleman, 1985, 1989, p. 57). In this way, such state actors increase their autonomy relative to the interest groups, a major aim of most political actors (Rhodes, 1981, p. 122; Laumann and Knoke, 1987, p. 13; Nordlinger, 1988, p. 882; Christensen and Christiansen, 1992, p. 7). State actors which cannot aggregate authority are weak and have little influence on the shapes of networks they attempt to design. Conflicting government agencies and interest groups, therefore, enter the network. The relationship between the participants tends to develop into an issue network which suffers from frequent and fundamental conflicts over policy making (Smith, 1993, p. 63). Policy communities are likely to emerge when state actors are in intermediate positions on the strong-weak state continuum. On the one hand, state actors cannot introduce a countervailing power into the network, while, on the other hand, they can exclude some interests. The state does not have a competing decision making centre with sufficient power to support the interest groups excluded. If a state actor attempts to introduce a countervailing power into the network, the insider interest group can mobilise resistance to this elsewhere in the state structure.

The causal link between state structures and meso-level policy networks is not, however, a strong one. Peculiar historical events may make their unique imprints on the shapes of networks. This means that the relationship between state structure and meso-level policy networks may be more or less in agreement with the one presented above.

Policy change in centralised and fragmented states

This section explores theoretically the way in which different organisational configurations of states (or political systems) either facilitate or constrain meso-level policy change when actors put the issue of policy change on to the agenda in sectors in which they do not traditionally participate in policy making.

Political actors (e.g. state actors or interest groups) attempting to bring about policy change in sectors in which they are not traditional participants require political power in order to be successful. They derive political power from their ability to generate authority, either by neutralising or minimising opposition. This can most easily be achieved in centralised states because they have well-developed coordination mechanisms which can be used to minimise the number of veto opportunities.[11] Actors who can generate authority have the capability to change the established order in a sector. To oppose such actors, those subject to policy change require veto opportunities, but, since there are only few of these in centralised states, fundamental policy change may occur. Another factor weakening status quo minded actors is that in centralised states, countervailing forces may have been introduced into the cores of meso-level networks earlier on with the result that strong coalitions opposing fundamental policy change are difficult to form (see chapter 2 on this point).

Political actors who pursue policy change in fragmented states have limited opportunities for generating authority. Since coordination is difficult, they are likely to meet opposition from one or more competing decision making centres within the state. This competition - whether it occurs between central and regional authorities, between actors at the same administrative level of the state or between the legislature and the executive - creates many veto opportunities for those wishing to maintain the status quo and causes deadlock within the state. This has the effect that actors can seldom generate sufficient authority to become strong reformers and, consequently, the opponents of change can prevent fundamental alterations from taking place. Even though there is somewhat of a tendency for fragmented states to develop issue networks which - due to their lack of cohesion - provide favourable opportunities for fundamental policy change (see chapter 2), deadlock within the state prevents actors from generating authority and, therefore, policy change is likely to be only moderate.

Moderately fragmented states are likely to produce outcomes similar to those of fragmented states. Although deadlock within the state is unlikely to occur, opportunities for change are limited because actors pursuing change cannot generate sufficient authority to gain control over the policy process. Since moderately fragmented states tend to develop cohesive meso-level policy communities which enable their members to create strong status quo orientated coalitions, the possibility for fundamental policy change is even further constrained.

The literature on the state has been the point of departure for the development of the macro-political model linking the broader organisational context and policy change. The model is developed at the level of the state but is also applicable at EC-level. As argued here, the way in which authority is distributed among institutions influences the opportunities for policy change. This is also true at EC level because the EC, like the state, consists also of a set of institutions which relate to each other in a certain way. This, in turn, has a major impact on the contents of policy change. Although the EC is not a state, it can be compared with states on the centralisation-fragmentation dimension. In other respects, it may make less sense to undertake a comparison.

Conclusion

In this chapter, it has been argued that the organisational structure of states (or political systems) and social groups' structural power in parliament affects the contents of policy change. Fundamental policy change is most likely to occur if the state is centralised and if the social group subject to policy change has limited structural power in parliament. However, these macro-political conditions are necessary but not always sufficient to provide actors pursuing fundamental change with political power. Policy change tends to be only moderate when the state is highly or moderately fragmented and the social group concerned has considerable structural power in parliament.

The aim of this chapter has been to identify macro-political variables which affect meso-level policy making. The purpose of integrating such variables is to explain policy outcomes in situations in which a political actor attempts to bring about policy changes in a sector in which it does not traditionally participate in policy making. To explain policy outcomes in such situations, we need to move beyond meso-level analysis. Realising that meso-level theoretical models do not provide satisfactory explanations alone, our analysis should integrate those features of the broader context which influence policy making at the meso-level. However, by so doing, we may lose sight of theory because we end up with too many independent variables. We can, however, limit this risk. In studying meso-level policy making, we implicitly assume that meso-level variables have most explanatory power. Therefore, the purpose of integrating macro-political variables is not to provide detailed explanation, but to outline the broader context within which meso-level policy making takes place. Keeping this in mind, we can avoid dealing with many macro-political variables all of which have some explanatory power.

Notes

1. See chapter 2 for a definition of structural power.

2. However, corporate pluralism, which had its heyday in Scandinavia in the late 1970s and early 1980s, did include MPs in the definitions of segments and sectors (e.g. Damgaard and Eliassen, 1978; Christensen and Egebjerg, 1979, Damgaard, 1981).

3. Until recently, the thesis had widespread support in Scandinavia (e.g. Damgaard, 1984; 1986; 1992, p. 85; Rommetvedt, 1992, p. 96). However, recent studies show that the Scandinavian parliaments are strong parliaments (Damgaard, 1992 ed., Damgaard, 1994).

4. In a parliamentary democracy, '... a government is composed in such a way that it is accepted by a majority in the assembly, or in one of its chambers, and that it leaves office if this is not found to be the case ... (Rasmussen, 1972, p. 225, quoted and translated in Damgaard, 1992, p. 14).

5. In a two-party system, we should devote more attention to factions within parties supporting the social group concerned.

6. If the party which is loyal to the social group examined is the dominant party, of course, this step of the analysis is part of the first two steps.

7. However, there is no guarantee that the parliamentary coalition will vote in favour of such groups' interests. Interest groups may not be able to take advantage of their structural power because they misjudge the political situation (see Marsh, 1994, p. 9).

8. Katzenstein also includes the differentiation between state and society as a variable that influences the strength of the state. A low degree of differentia-

tion produces strong states (1978, pp. 314, 323).

9. For example, Katzenstein (1978) argues that the German state is stronger than the British.

10. Authority refers to the right to make decisions within a policy area.

11. It must be said, however, that we cannot take for granted that a centralised state always produces unitary action; but in comparison with a fragmented one, the likelihood is greater.

4 Classifying policy reforms and new policies

Introduction

Chapters 2 and 3 linked policy networks and the broader political context to the nature of policy change. The two chapters distinguished between moderate and fundamental policy change. In the former type, established policy principles underpin the policy choice. In fundamental change, new policy principles are introduced and become the basis of the choice of policy. A policy principle defines what should be the basic idea of public policies and outlines, in general terms, which solutions are attractive and feasible.

As already stated in the introductory chapter, the concept of policy change includes both the reform of existing policies and the introduction of new policies in a sector or sub-sector. New policies are introduced to cope with problems established policies do not address. Both types of policy change threatens the established sectoral or sub-sectoral order. In chapter 1, the literature on policy networks was criticised for putting too little emphasis on the conceptualisation and classification of policy outcomes. Here, I attempt to remedy that shortcoming by classifying theoretically the contents of new policies and reformed policies. The former are classified into high and low cost policies and the latter into first, second and third order reforms. This chapter links policy networks and the broader political structures with specific policy choices in reform processes and in processes in which new policies are introduced. In order to do so, we must develop classifications of policy contents which are logically related to the characteristics of various network types. The concept of policy principles enables us to link the characteristics of various policy network types and broader political structures with the characteristics of different types of new policies and policy reforms. As argued above, the most radical types of policy changes involve a change of policy principles. Policy change is possible when policy network and broader political structures do *not* privilege the interests of those favouring the maintenence of established policy principles as the basis of policy change.

Before the classifications of new policies and policy reforms can be developed, it is necessary to account for what constitutes a public policy. It is argued in the section to follow that a public policy consists of objectives and of instruments to achieve these. Instruments and objectives are, thus, important components of the classifications of new policies and policy reforms. The next step is to link the classifications with various types of policy networks and broader political structures. The linkage is formulated as a theoretical proposition which is then transformed into an empirical to be used in the comparisons of agri-environmental policy making and agricultural policy reform.

Policy principles rarely reveal themselves to the researcher. This is particularly a problem in the empirical classification of new policies; policy reforms can be classified without detailed analysis of the policy instruments applied. The best way to reveal the principles upon which a new policy is based is to apply the policy's instruments as indicators (see chapter 5 on this point). We, therefore, need to consider which instruments indicate whether high or low cost policies have been employed. This chapter develops operational definitions and classifications of environmental policy instruments in order to undertake a comparison of agri-environmental policies in Denmark and Sweden.

What constitutes a public policy?

The concept of public policy refers to the content of state intervention. Basically, a public policy consists of two components. The first component concerns policy objectives. They can be formulated in broad and imprecise terms, or they may be specific and precise. Sometimes they are not even explicitly stated. Policy instruments are the second component of public policies. A policy can consist of a single instrument or a set of instruments employed to achieve one or several objectives.

Policy makers view the choice of policy instruments as more important than the choice of policy objectives (Sabatier, 1986, p. 29; Vedung 1997) because instruments have immediate consequences, while a policy's objectives 'are something to be achieved in the future' (Doern and Phidd, 1983, p. 112). Hedrén (1993, p. 32) clearly shows empirically that the settlement of policy objectives has limited importance in the policy process. He reports that Swedish members of parliament agree that conflicts over the settlement of policy objectives in environmental policy making are rare, whereas the choice of policy instruments more often leads to disagreements. Usually, politicians are more concerned with short rather than long term decisions. Policy objectives can be achieved by the use of various instruments and since the instruments have immediate consequences, they tend to become the centre of political dispute rather than the objectives.

However, these findings should not lead to a research approach in which we ascribe no importance to the settlement of policy objectives. Firstly, they have symbolic meaning in relation to the public; the main function of the objectives may be to tell the public that something is being done about a problem. Secondly, policy objectives may serve as a resource for implementing officials; they can derive some authority from the official objectives when trying to have actors fulfil desired aims. Thirdly, policy goals may have a function in policy evaluation (Sabatier and Mazmanian, 1981, p. 10; Sabatier, 1986, p. 23).

Instrument analysts have defined policy instruments in various ways. For instance, Salamon and Lund (1989, p. 29) define a policy instrument 'as a method through which government seeks a policy objective.' This definition is far too broad. Instead, a policy instrument should be defined as a political method which aims at bringing about a certain behaviour among individual members of the target group through learning, compulsion or economic incentives (Daugbjerg, 1995, p. 34). The choice of policy instruments is not merely a technical problem left to bureaucrats or experts but a highly political process (e.g. Daugbjerg, 1993; 1995; Doern and Phidd, 1983, p. 111; Howlett, 1991; Jansen and Lundqvist, 1993, p. 14; Linder and Peters, 1989, p. 36; Majone, 1989, p. 118; Howlett and Ramesh, 1994; Salamon 1981; Winter, 1994). Elmore (1987, p. 175) suggests that the choice of policy instruments 'has more to do with coalition politics than with their operating characteristics.'

Policy instruments and objectives cannot be analysed in isolation from the way in which policy makers organise the implementation process. Those who are to implement policy have an influence upon the instruments and goals which policy makers adopt. Those who are subject to regulation may, on the one hand, be willing to accept relatively tough objectives and policy instruments if a friendly government organisation obtains programme responsibility. On the other hand, if the organisation is non-friendly, tough objectives and instruments tend to be strongly opposed (Daugbjerg, 1995a). When we assess the political consequences of specific choices of instruments and objectives, we need to consider how the choice of implementation structure influences the application of the instruments when delivered to the individual member of the target group. Focusing on the instruments and objectives alone may reveal only part of the picture because some organisational forms strengthen their effects, while others weaken them. Likewise, the seriousness with which implementing officials take the official policy objectives also depends on the specific form of policy organisation. Therefore, the implementation structure is a conditional variable in the implementation process, influencing the effects of policy instruments and objectives. I do not develop a theoretical model for the choice of implementation structure, but I do consider implementation structures in the empirical analysis.

Classifications of new policies and policy reforms

This section firstly classifies new policies designed to cope with problems which an established policy does not address into high and low cost policies. Secondly, it classifies reforms of established policies into first, second and third order reforms. The two types of policy change can be more or less radical. The most radical ones introduce new policy principles into an established policy sector whereas the least radical base the choice of policy instruments and objectives on existing sectoral policy principles.

70

The least radical type of *new policy* is a low cost policy. Costs refers to the way in which those being regulated perceive the consequences of the policy. A low cost policy is characterised by the transfer of the regulated sector's established policy principle to the new policy, for instance the transfer of the major principle of agricultural policies - the principle of state responsibility - to agri-environmental policy making. Policy principles define the types of policy options which are attractive and feasible. When members of a network perceive established policy principles as socially, economically and politically beneficial, they attempt to transfer these principles to new policies in their policy sector. By so doing, they can minimise the perceived political and economic costs of the new policy. For example, the transfer of the principle of state responsibility from agricultural policies to agri-environmental policies implies that the state, and not the farmers, will pay a large share of the economic costs of pollution control and also that agricultural interests are taken into account in the choice of policy. The use of established policy principles means that the policy instruments applied limit the economic costs of regulation. Furthermore, the instruments give the group subject to regulation a certain control over the implementation process and over future policy formulation. The objectives of low cost policies do not, in practice, contradict the objectives of the established sectoral policy.

Table 4.1
New policies: low and high cost policies

Policy type	Characteristics
Low cost policy	Transfer of policy principles from the established policy to the new policy. Policy instruments pass political and economic policy costs on to groups other than those being regulated. Policy objectives do not contradict established sectoral policy objectives
High cost policy	New policy principles. Policy instruments concentrate policy costs on the groups regulated. Policy objectives contradict established sectoral policy objectives

The most radical type of new policy is a high cost policy. It is based on principles which differ fundamentally from the established principles of the sector concerned. New policy principles bring about policies which distribute social, political and economic costs and benefits fundamentally different from the existing one. While the established policy provides social, political and economic benefits to the group it regulates, the new policy inflicts cost upon it. The example of agriculture illustrates this point. The state and consumers bear the costs of regulation in agricultural policies. The employment of the polluter pays principle in agri-environmental policy making means that farmers will bear the economic and political costs of pollution control. Since established policy principles are not transferred, policy instruments which concentrate political and economic costs on the group to be regulated may be employed. Moreover, the instruments do not allow the group subject to regulation much control over implementation and over future policy formulation. Policy objectives contradict the established sectoral policy objectives.

Policy reforms can also be more or less radical. Peter Hall (1993, pp. 278-9) distinguishes between first, second and third order changes. In order not to confuse the termonology applied here with Hall's, the terms first, second and third order reforms are used to describe the three types of policy change which Hall describes. In first order reforms 'instrument settings [or levels] are changed ... while the overall goals and instruments of a policy remain the same.' Second order reforms have taken place 'when the instruments of a policy as well as their settings are altered [and] ... the overall goals of a policy remain the same.' He characterises third order reforms by alterations in 'the instrument settings, the instruments themselves, and the hierarchy of goals behind policy.' However, it is doubtful whether a change in the hierarchy of goals involves a third order reform. In my view, a shift in policy goals involves no more than a second order reform. It is unclear from Hall's work what third order reforms actually involve because he argues that policy reforms which change the ideas - or policy principles (the term applied in this book) - on which a policy is based, involve a 'paradigm shift' (ibid., p. 279). Thus, there seems to be a contradiction in Hall's categorisation of a third order reform and what characterises it. A change in the hierarchy of policy goals does not involve a 'paradigm shift' since a policy is not radically changed if a goal which has previously been downgraded gets priority. Hall would probably agree that it does not involve a third order reform if, for instance, agricultural policy makers upgrade consumer concerns[1] and downgrade concern for farmers' income and the question of subsidising farmers is not even discussed. My suggestion is to characterise policy reforms as follows:

Table 4.2
Policy reforms: first, second and third order reforms

Reform type	Characteristics
First order reform	Instrument settings, or levels, are changed. Objectives, instruments and policy principles remain the same.
Second order reform	Objectives, instruments and their settings are altered. Policy principles remain the same.
Third order reform	Objectives, instrument settings, instruments and policy principles are changed.

Theoretical and empirical propositions

In this book, policy change is defined as reform of existing policies or introduction of new policies designed to cope with problems an established policy does not address. The above classifications of the two types of policy change are constructed in a way which enables us to link them to the policy network model developed in chapter 2 and the macro-political model developed in chapter 3. These models distinguished between moderate policy change in which existing policy principles remain the basis of policy choices and fundamental change in which new policy principles are introduced to underpin the choice of policy. The above classifications of new policies and policy reforms enable us to distinguish between fundamental and moderate policy change. Thus, in the language of the classifications, fundamental policy change involves the introduction of high cost policies and third order policy reforms because new policy principles are introduced. Moderate policy change is associated with low cost policies and first or second order policy reforms since these are based on existing policy principles.

The theoretical models developed in chapters 2 and 3 suggest how policy networks and the broader political context are linked to the nature of policy change when actors attempt to bring about policy change in sectors in which they do not traditionally participate in policy making. It has been argued that the most favourable opportunity for fundamental policy change is most likely to occur when a non-cohesive policy network exists in the sector in which change is on the agenda, when the state is centralised and when the parliamentary support of the group subject to policy change is limited. In such a situation, it is unlikely that this group can mobilise sufficient support to defend the status quo successfully and therefore new high cost policies and policy reforms involving third order reforms tend to be adopted.

Moderate policy change tends to occur when the established meso-level policy network is cohesive, when the state is fragmented and when the group facing pressure for change has a high degree of support in parliament. These conditions enable status quo minded actors to mobilise sufficient support to prevent fundamental policy change, but since they are rarely able totally to dismiss demands for changes, they cannot avoid making some alterations in policy. These are moderate, meaning that new policies are low cost policies and that policy reforms involve only first or second order reforms.

Ideally, an empirical test of a theoretical model would test all the elements of the model, but such an analysis would demand many resources and require high quality data which is very difficult, if not impossible, to obtain. Usually, we cannot show all causal links in a policy process. The second best approach is to develop strategic propositions which test the overall validity of the model. Such a design relies on data on dependent and independent variables. If the empirical relationship between both types of variables is as predicted by the theoretical model, then we have strong reasons to believe that the theoretical model is valid. A good theoretical model accounts for all the links between the independent and the dependent variables.

In the analysis of agri-environmental policy making and agricultural policy reforms, the strategic proposition below enables us to assess the utility of both the meso and the macro-level theoretical model developed in chapters 2 and 3 respectively. The empirical proposition suggests:

> Policy makers have good opportunities to adopt high cost agri-environmental policies and third order agricultural policy reforms if a non-cohesive policy network exists in the agricultural sector, if farmers' structural power in parliament has declined and if the state is centralised. And policy makers tend to adopt low cost agri-environmental policies and first and second order agricultural policy reforms if the agricultural policy network is cohesive, if farmers have maintained structural power in parliament and if the state is fragmented.

An empirical analysis producing findings which fit the proposition strengthens the validity of the theoretical models. However, if I find third policy reforms and high cost policies where cohesive agricultural policy networks exists, where farmers have structural power in parliament, and where the state is fragmented, or vice versa, then the theoretical models are not valid. If policy makers adopt a third order policy reform or a high cost policy in an agricultural sector with a non-cohesive policy network and farmers have structural power in parliament and/or the state is fragmented, then the articulation between the meso-level and the macro-level is not valid. In such cases, the characteristics of the policy network explain the choice of policy. Nor is the meso-macro link valid if a third order policy reform or a high cost policy occurs in an agricultural sector with a cohesive network and farmers have a low degree of structural power in parliament and/or the state is centralised. Such a finding implies that it is not network characteristics but macro-political factors which explain policy choices.

Developing operational definitions of environmental policy instruments

So far, it has been argued in this chapter that low cost policies consist of policy instruments which pass political and economic policy costs on to groups other than those being regulated. High cost policies concentrate the costs on those who are regulated. To utilise the concepts of low and high cost policies in empirical research, we need to develop operational definitions establishing those instruments which concentrate the political and economic costs on the groups regulated and those which pass them on to other groups. In applying the two concepts in the study of agri-environmental policy making, it is necessary to clarify which instruments farmers perceive as respectively high and low cost instruments.[2]

The literature on policy instruments has produced a large number of typologies and classifications. Some of them operate only with few categories or types, whereas others have many.[3] However, these typologies and classifications cannot be applied in this book because they reflect the concerns of those who develop them rather than of those who choose the instruments. This is one reason why research has evolved many different typologies and categorisations. Jansen and Lundqvist (1993, p. 1) have reviewed the literature on policy instruments and find that there seem to be just as many classifications as there are interpreters. In this section, I point out the problems in some policy instrument typologies. Further, I attempt to develop a operational definition which is grounded in the logic which characterises the policy formulation process.

Political actors' perceptions of specific policy instruments may be different from those of the researcher. The perception of the latter should not, therefore, be the basis of an instrument typology if our purpose is to explain the choice of instruments. As Linders and Peters state: 'It is the policymaker's biases - not the investigator's - that shape the relative influence of context and of the perceived attributes of instruments on design criteria' (1989, p. 55). What is important is how the participants in the policy process perceive which policy options are costly or beneficial to them (Wilson, 1980, p. 366). There need not be any connection between the actual costs of a policy instrument and different groups' perception of its costs (Daugbjerg, 1995, p. 36).

Policy instruments can be ordered along several different dimensions (Salamon and Lund, 1989, p. 31). Some base their classifications on the 'authoritative force involved in the governance efforts' (Vedung, 1997) or, in other words, the degree of coercion (Linders and Peters, 1989, p. 46; Doern and Wilson, 1974, p. 339; Doern and Phidd, 1983). Another group of policy analysts apply categorisations without specifying the foundations on which they were developed (for examples, see Linder and Peters, 1989, p. 44; Salamon and Lund, 1989, pp. 31-5). Those basing instrument typologies on the theme of coercion argue that governments, for reasons of legitimacy, prefer to use the least coercive instruments first and apply more coercive ones later (Doern and Wilson, 1974, p. 339; Doern and Phidd, 1983, p. 128; Vedung, 1997). Vedung (1997) develops a threefold taxonomy which classifies instruments according to how coercive they are. The three categories of his taxonomy are, ranked from the least to the most coercive category, informative, economic and regulatory instruments (see also Vedung, 1991, ch. 7). Vedung's classification suggests that regulations are more coercive than economic instruments because the former set out mandatory requirements, whereas the latter neither prohibit nor prescribe particular actions. Economic instruments just change the incentive structure of those being regulated and, thus, provide more choice opportunities than regulations (Vedung, 1997). Informative instruments are least coercive as they involve no obligations and 'no handing out or taking away of material resources' (ibid.).

The great advantages of the threefold classification are its simplicity and parsimony and the fact that it makes sense for policy makers (Jansen and Lundqvist, 1993, p. 5). However, the underlying logic of the classification is problematic. If we follow the logic of Vedung's classification, those subject to regulation would prefer the least coercive instruments to the most coercive. Consequently, they would prefer economic instruments to regulations and information to economic instruments. This argument does not necessarily accord with the real world. In environmental policies, for instance, interest groups often prefer regulations to economic instruments although the latter provide many more opportunities for choice than the former. Therefore, there must be another logic involved in the instrument choice than the one Vedung suggests.[4] Consequently, an operational definition needs to be based on another underlying logic.

Most policy instrument analysts base their classifications on their own concerns rather than those of policy makers. For that reason, we have only limited possibilities to achieve an insight into the choice of policy instruments. In order to explain this, we need to reveal how policy makers conceive of various policy instruments. As Steinberger (1988, p. 189) points out: 'the best of the typologies are plausible and useful in that they describe (or, rather, can be used to describe) *typifications* that are generally and commonly employed by participants in the political process to define public policies.' Or in Linder and Peter's (1989, pp. 54-5) words: 'It is the policy-maker's criteria - not the investigator's - that structure judgements about instruments and serve as the basis for any working taxomony.' It is important to point out that different types of participants in the policy process have different perceptions of policy instruments (Steinberger, 1980, p. 190). Consequently, we have to make an explicit choice as to whose perception should be the basis of the operational definition. The operational definition being developed in this chapter is based upon the presumed perception of farmers.

We can only apply the operational definition in industries in which there is perfect competition. An industry is characterised by perfect competition when each individual market participant is unable to influence market prices. Such a market has many, relatively small producers who produce similar products. Each producer has a market share which is far too small to influence market prices (Nedergaard, 1988, p. 27). It does not matter to them whether markets or political decisions set prices. Since political decisions set prices on agricultural commodities, farmers are price takers who do not compete with each other (Daugbjerg, 1994b, p. 86).

Above, a policy instrument was defined as a political method which aims at bringing about certain behaviour among individual members of the target group through learning, compulsion or economic incentives. This definition suggests that policy instruments are based on three types of inducements, namely economic, legal or voluntary inducements. Economic inducements work in two different ways: they can be positive or negative. As Vedung (1997) argues: 'In the positive case, a material resource is handed over to the agent, whereas in the negative the individual or organization is deprived of some material resource.' Thus, positive inducements reward certain behaviour, typically by using subsidies or tax reductions. Negative inducements make an unwanted activity more costly, typically by the use of taxes, charges or fees. It is important to distinguish between the two types of economic inducements because they produce different distributive outcomes. We can sub-divide legal inducements into prohibitions and prescriptions (Vedung, 1991, p. 92; 1997; Jansen and Lundqvist, 1993, pp. 9-13). However, these two types do not produce significantly different distributive outcomes and, consequently, there is no reason to sub-divide legal inducements, complicating the operational definition more

76

than necessary. So, in classifying policy instruments on their inducements, we can distinguish between:

- positive economic instruments
- negative economic instruments
- regulatory instruments
- informative instruments

These instrument types produce different political and economic costs for farmers. Negative economic and regulatory instruments are the two types which farmers tend to perceive as the most costly. On their own, positive economic instruments are beneficial; however, they are often used to decrease the economic costs of implementing regulatory instruments. Informative instruments produce benefits because voluntarism allows farmers to ignore advice if they think it will impose costs upon them. Therefore, they only follow advice when they believe they benefit from it.

In principle, environmental policies concentrate the economic costs of state intervention upon polluters. Concentrated costs are very visible (Leone, 1986) and thus controversial. Therefore, the distribution of costs becomes the main theme in the environmental policy process. Costs motivate farmers to mobilise politically. The distribution of cost is much more important for an interest group than the sharing out of benefits (provided all members gain). It is more painful to lose than not to get something one does not have (Christiansen, 1990, p. 446). Hence, in environmental policy making, interest groups listen carefully to the demands of the rank-and-file members in order to avoid unrest within the association. Dissatisfied members may find it much easier to mobilise opposition against burdensome state intervention than to mobilise support for more benefits (see ibid). In environmental policy making, interest groups representing those subject to regulation, therefore, have a great incentive to ensure that a large majority of their members feel that costs are fairly distributed. By contrast, policy instruments providing benefits make all members better off than before, although some get relatively more than others (Daugbjerg, 1995, p. 36). Environmental policy also involves political costs because it as a new policy brings new participants into the policy process. Traditional participants see such integration as a political disadvantage if these new actors achieve control over the new policy or if the instruments of the policy keep traditional actors away from influence on the instruments' settings. Because political actors are very aware of economic and political costs, it is necessary to take a closer look at both regulatory and negative economic instruments.

We can assume that farmers perceive negative economic instruments as more costly than regulatory instruments. The latter have the lowest political costs. From the point of view of interest associations, negative economic instruments, e.g. green taxes, have the highest political costs because they create future uncertainty for polluters. In general, interest groups have limited control over the making of tax policies (see Damgaard and Eliassen, 1978, pp. 302-11; Damgaard, 1986, pp. 275, 284). Therefore, they are not in a position to block future increases in the tax level and this creates uncertainty. By contrast, regulatory instruments provide much better opportunities for polluters to achieve control over the future policy process and therefore certainty. To implement such instruments, public authorities require information and knowledge which interest groups possess (Bressers, 1995, p. 13). Further, to avoid implementation gaps, it is necessary for an implementing agency to obtain support from interest groups representing polluters. Both factors imply that polluters achieve a certain control over the policy process, which they cannot achieve in tax policy making.

Moreover, the costs of negative economic instruments are much more visible than those of regulatory instruments. The cost of a pollution charge or another type of green tax shows up on the balance sheet, whereas it is much more difficult to estimate the costs of complying with regulatory measures. For an interest association, there is a risk that visible costs will lead its members to think that it did not assert itself sufficiently in the policy formulation process and, thus, give rise to dissatisfaction within the association. No association wants this to happen.

At the enterprise level, regulatory instruments also have more advantages than negative economic instruments. They often allow negotiations between environmental inspectors and polluters. In negotiations concerning specific implementation, polluters can take advantage of having more information and knowledge on actual production processes than environmental inspectors. This means that polluters may be able to persuade environmental inspectors to accept those solutions which are the cheapest to implement.

In economic terms, the two types of instruments also differ. It can be assumed that agricultural organisations think that the economic costs of compliance for individual farmers are much higher for negative economic instruments than for regulatory instruments. Taxes on pollution (e.g. effluents or discharges) or on production inputs which cause pollution are paid on all units of pollution or on all units of the polluting production input applied. For instance, a tax on fertilisers means that a farmer pays tax for all units of fertiliser used to produce cereals. Furthermore, a tax will reduce benefits because higher production costs force a producer to decrease output. Thus, s/he loses the profit of the part of production which was profitable before the tax was introduced but is now unprofitable. The fertiliser tax illustrates this argument. It increases the costs of each unit of cereals produced, and therefore the farmer will use less of it because the costs of applying an extra kilo of fertiliser equal the benefits of using that kilo at a lower level of production than before the tax was introduced. Since the farmer had a profit on the production exceeding the new optimal production level before the introduction of the tax, s/he now loses income as a result of the lower level of production.

Regulatory instruments need not by definition be cheapest to implement because implementation may require large investments. Usually, however, they are cheaper for polluters because they tend to cause fewer profit losses. If a regulatory instrument sets a standard (or a non-tradeable quota) for pollution (or polluting inputs), a producer loses only the benefits of the production which can no longer be produced because of the standard (Svendsen, 1996, p. 103). The fertiliser example can also illustrate this point. If a quota is applied, the production costs are the same as before it was introduced, but the farmer will not be able to use fertilisers up the point where the costs of adding an extra kilo fertiliser equal the benefits of it. Hence, s/he will lose part of the profits earned before the quota was applied. Thus, a regulatory instrument brings about fewer losses in profits than a tax because as well as losing the profits brought about by decreased production, the tax also puts an extra cost on the profitable production. This does not happen when a regulatory instrument is used.

Besides being based on inducements, environmental policy instruments involve different modes of adjusting instruments to polluters' specific conditions. An instrument can be adjusted to individuals or enterprises in three different ways. At one extreme, the degree of individual adjustment is low: that is, the instrument is used in the same manner for all members of the group being regulated: in other words, it is universal. A simple and fixed tax on fertilisers is a good example. The tax is not adjusted to individual conditions, for instance the quantity of animal manure or the choice of crops. At the other extreme, the degree of individual adjustment is high because the instrument is adjusted to the specific conditions of each individual or firm. An example is the Danish system of obligatory fertiliser and crop rotation plans which aims at limiting nitrate run-offs in agriculture. The instrument requires each farmer to work out a plan which takes account of specific production. The plan sets binding prescriptions for the application of manure and fertilisers. In between these two extremes, instruments can be adjusted to specific areas, groups or types of production. In Sweden, for instance, the requirement on autumn crop cover relates to specific areas; in Southern Sweden, farmers are required to have a larger share of autumn crop cover than their colleagues in Northern Sweden. Furthermore, in both Sweden and Denmark some instruments are adjusted for the specific type of production. For example, the requirements on storage capacity for manure are different, depending on whether farmers have pigs or cattle.

It can be assumed that farmers perceive universal policy instruments as more costly than individually adjusted instruments because the former 'are generally inaccurate and error-prone; that is, they may mis-define subjects or misdirect them' (Linder and Peters, 1989, p. 46). Individually adjusted instruments, by contrast, are much more precise (ibid.). Due to their inaccuracy, universal policy instruments tend to produce effects which farmers view as unfair. The costs of such instruments may be distributed in unintended ways. Individually adjusted policy instruments involve much lower risks of unintended consequences; they, therefore, are more acceptable to farmers. In pollution control, individualised instruments distribute costs in a way which inflicts the greatest costs on those who pollute most, whereas small polluters will only face limited costs. Farmers perceive such a situation as fair. A survey carried out by Bager and Søgaard (1994, p. 80) concludes that most Danish farmers assess their consumption of fertilisers to be below average; only a very small percentage assesses it to be higher. These results indicate that the vast majority of Danish farmers believe that they pollute less than average. Therefore, individualised instruments seem fair to them because those who cause the pollution pay the costs of pollution control.

The competitive characteristics of markets influence the perceptions of individually adjusted instruments. While such instruments may be problematic in markets characterised by imperfect competition (Leone, 1986, p. 17), they cause no problems for producers in markets with perfect competition..Indeed, the character of the market is the reason we can assume that farmers perceive the costs of individually adjusted instruments as lower than those of universal instruments. Farmers are price takers; thus, they do not fear competition from other farmers. They are, therefore, much more concerned with their own costs of compliance rather than with those of others. Farmer A need not fear that farmer B shapes the delivery of policy in a way which limits B's costs of compliance. Different costs do not affect the market situation of farmer A. Indeed, it may be an advantage for all farmers if some of them can limit the costs of compliance. The absence of competition among farmers implies that those who succeed in reducing costs have no reasons to keep their knowledge secret.

The complexity of individually adjusted instruments is another reason they may be more acceptable to farmers than universal instruments. Complex instruments leave some room for flexibility as implementation officials cannot check whether the individual farmer complies strictly with the rules. Consequently, farmers can avoid some of the instruments' intended effects, but it must be said that experienced interest groups see members' refusal to implement instruments as unwise. Such groups do not act independently of the context surrounding them. They have allies (local communities, government organisations, politicians, etc.) on whom they rely for support in public policy making. Ignoring the demands of these allies could bring about new and costly policy measures in the future (Mitnick, 1993, pp. 69-70). Attempts to get around the implementation may damage an interest group's political credibility. Therefore, experienced interest groups will try to persuade their members to comply with policy instruments having low perceived costs.

Basing the operational definition upon deduction, we can assume that in environmental policies:

- Farmers prefer positive economic policy instruments and informative instruments to regulatory instruments which, in turn, are preferable to negative economic instruments.
- Farmers prefer policy instruments which are adjusted to individual conditions to those which are adjusted to regional, group or production type conditions. The least desired are universal instruments.

Brief outline of the case studies

This book raises a number of questions which empirical analysis must answer. Firstly, was the nature of the nitrate policies adopted in Sweden and Denmark fundamentally different from the nature of established agricultural policies? Secondly, were the agricultural policy reforms in the EC and in Sweden fundamental policy changes or were they merely adjustments of the existing policy? After the compari-

80

son of nitrate policies and agricultural policy reforms in chapter 5 and chapter 6, the book, in chapter 7, proceeds to examine whether the cohesion of agricultural policy networks can explain differences in policy choices. Finally, in chapter 8, a macro-level analysis is carried out. It attempts to explain whether differences in farmers' structural power in parliament and in the organisational structure of states contribute to explaining the choice of policy.

Notes

1. For example, by emphasising that consumers have access to sufficient supplies of quality foods at reasonable prices.

2. The operational definitions are employed in the comparison of Swedish and Danish nitrate policies. The purpose of the comparisons is to assess which of the policies share most features of respectively a high cost and a low cost policy. Hence, we do not need a detailed rank ordering which shows the perceived political and economic costs and benefits of various policy instruments and of various combinations of instruments.

3. The number of categories varies from two to 63 (Vedung, 1997).

4. See also Howlett and Ramesh (1994, pp. 7-8) and Linders and Peters (1989, p. 47) for a critique of coercion as the basis of instrument typologies.

5 Danish and Swedish nitrate policies

Introduction

Sweden covers an area ten times larger than Denmark.[1] While the Danish landscape is dominated by cultivated farm land (64 per cent), forests predominate in Sweden (60 per cent). Despite these differences, there are several common features in the structure of agriculture. The area of cultivated land in Denmark and Sweden is about the same size: slightly less than three million hectares. The number of farms is not significantly different: in 1992, there were 91,850 farms in Sweden of which 38 per cent were run by full-time farmers; in Denmark, the figures are 74,460 and 46 per cent respectively. 3 per cent of the Swedish labour force work in primary agricultural production; in Denmark, it is a little less than 5 per cent.

The comparison focuses on the making of national nitrate policies as an example of agri-environmental policy making because Danish and Swedish policy makers paid most attention to the contamination of ground and surface waters by nitrate (see Eckerberg, 1994; Andersen and Daugbjerg, 1994). As was pointed out in the introductory chapter, Danish nitrate policy is different from Swedish. Although farmers in Denmark experience climatic conditions which are significantly different from those of farmers in Northern Sweden, the nitrate policies of the two countries can be compared without major problems. Firstly, the climate in the intensive farming areas in Southern Sweden is similar to the Danish. Secondly, the two countries produce the same types of agricultural products which, to a large extent, cause similar pollution problems. Thus, differences in climate do not account for the variation in the contents of the nitrate policies. Nor can differences in scientific knowledge on nitrate pollution provide an explanation. The exchange of knowledge and information between the scientific communities of the two countries is well developed. Swedish researchers often quote Danish reports and vice versa. Furthermore, Nordic networks in nitrate research have been established (see e.g. Nordic Council of Ministers, 1992; 1992a). Since none of these factors can explain why the nitrate policies differ, this chapter turns to political factors.

One purpose of this chapter is to examine the contents of Danish and Swedish nitrate policy. How do they differ? In chapter 4, a classification which distinguishes between high cost and low cost policies was developed. This can be used to classify nitrate policies. In a low cost nitrate policy, the major principle of agricultural policy making, the principle of state responsibility, is transferred from the agricultural policy to the nitrate policy. Basically, the application of that principle implies that the state takes care of the well-being of farmers. Applied in nitrate policy making, the principle would ensure that the nitrate policy does not inflict too many and too high political and economic costs on farmers. When the principle of state responsibility underpins the choice of policy instruments and objectives, the nitrate policy will consist of instruments passing costs on to groups other than farmers and objectives which do not contradict the objectives of the agricultural policy. A high cost nitrate policy is based upon the polluter pays principle. Unlike the principle of state responsibility, the polluter pays principle does not protect farmers from the costs. Thus, a high cost policy employs instruments which put costs on farmers and its objectives may contradict the objectives of the agricultural policy.

Since the policy principles upon which a policy is based are rarely explicit, the best method to classify Danish and Swedish nitrate policy is to analyse the use of policy instruments (Daugbjerg, 1995b, 1998). Chapter 4 developed the operational definitions of environmental policy instruments, concluding that we can assume that farmers prefer positive economic and informative instruments to regulatory instruments. Negative economic instruments are the most costly seen from the farmers' point of view. Moreover, there are good reasons to assume that farmers believe that the more policy instruments are adjusted to individual conditions, the less disadvantageous they become. The choice of policy objectives also help to indicate the type of nitrate policy. Objectives which are formulated in general terms indicate that the nitrate policy is a low cost policy. Such objectives do not contradict agricultural policies. An important objective in agricultural policies is to safeguard farmers' income. Precise and specific objectives may, at the end of the day, affect farmers' income because the fulfilment of such goals may be costly. Therefore, they are in conflict with agricultural policies. By analysing the choice of policy instruments and objectives, this chapter establishes whether Danish and Swedish nitrate policies are high cost or low cost policies.

Agri-environmental policy processes are characterised by conflicts over which policy principles should underpin the choice of policy. While farmers prefer the principle of state responsibility, environmental interests favour the polluter pays principle (see the introductory chapter). Chapter 2 argues that a crucial factor influencing the choice of policy principles is the policy positions of the state actors most centrally positioned in the established sectoral policy network into which a proposal on a new policy is introduced. Therefore, empirical analysis of pollution control in agriculture must focus on the agricultural state actors' policy preferences. How do they position themselves in relation to agricultural and environmental interests? If they ally with farmers, the most likely outcome of the process is a low cost policy. By contrast, when agricultural state actors pursue their own bureaucratic interests or try to mediate between environmental and agricultural interests, high cost nitrate policies are likely to be chosen. Chapter 2 also points out that the type of the agricultural policy network is an important variable explaining the policy preferences of agricultural state actors. I shall return to this issue in chapter 7.

The organisation of the state administration in Denmark and Sweden is not similar, thus implying that the configuration of state actors in agricultural policy making is different. In general, Danish government departments are the major state actors in policy formulation. Agencies (or directorates) are mainly concerned with administration and rarely achieve a central position in the policy process. Environmental policy making does not, however, fit this pattern. The Environmental Protection Agency (EPA), which is attached to the Ministry of the Environment, has a unique and strong position in the environmental policy process. It carries out functions for which departments in the other ministries are responsible. These functions include negotiations with organised interests in public policy formulation (Pedersen and Geckler, 1987).

Swedish agencies generally have a role in policy making which is similar to that of the Danish Environmental Protection Agency. According to the Swedish constitution, the *Riksdag* (the Swedish parliament) cannot influence the administration of the laws it enacts. Consequently, the implementing agencies are much more independent than in most other Western countries. In practice, there are, however, fairly close contacts between the government departments and the agencies (Petersson and Söderlind, 1993, pp. 73-9). Compared to other countries, the staffs of Swedish government departments are small with the result that usually they have to turn to the agencies for bureaucratic resources. The independent status of the agencies and the departments' needs for bureaucratic resources bring the agencies into a central position in the policy process. In Swedish agricultural policy making the Agricultural Marketing Board (*Statens Jordbruksnämnd*) played an important role in the agricultural price policy. The National Agricultural Board (*Lantbruksstyrelsen*) had a central position in the agricultural structural policy, in the agri-environmental policy and in providing advisory service to farmers. In 1990, the two agencies were merged into one agency: the Swedish Board of Agriculture.

Farmers have organised differently in the two countries. Since 1971, Swedish farmers have united their union and cooperative interests in a single association: the Federation of Swedish Farmers (*Lantbrukarnas Riksförbund, LRF*). Usually, it is referred to by its official abbreviation, LRF. Danish farmers have formed three associations. The Danish Farmers' Union (*De danske Landboforeninger*) represents medium-sized and larger farms, the Smallholders' Union (*Danske Husmandsforeninger*)[2] represents small farmers and the Association of Commercial Farmers (*Dansk Erhvervsjordbrug*) organised the largest farms until 1996 when its was merged with the Farmers Union. Clearly, the Farmers' Union and the Smallholders' Union exerted the most influence. Both were, and still are, attached to the Council of Agriculture (*Landbrugsraadet*) which represents both union and cooperative interests in Danish agriculture. The Association of Commercial Farmers has had only minor influence in agricultural and agri-environmental policy making presumably because it organised less than 2 per cent of the farmers.

Denmark

During the 1960s, the public in most Western countries, including Denmark, became increasingly concerned about the state of the natural environment. This motivated politicians to intervene into the issue area and enact environmental laws and set up agencies to implement them. Denmark was no exception. The preparation of the Environmental Protection Law began in the late 1960s and was finished in 1973.

In 1969, the Danish government set up the Pollution Board to investigate environmental problems and to suggest policy measures to diminish pollution. The Board also addressed agricultural nitrate pollution. It concluded that farmers' application of nitrogen could cause local pollution problems and that the increasing concentration of animal husbandry and the decreasing value of manure, compared to chemical fertilisers, might exacerbate problems in the future. To cope with agricultural pollution, the Board recommended that large animal farms should apply for pollution permits (Forureningsrådet, 1971, p. 139).

The Pollution Board produced 29 reports which became central in the making of the Environmental Protection Law in 1973 (Andersen and Hansen, 1991, pp. 37-8). In accordance with Danish political traditions, the Ministry of the Environment[3] negotiated with the organised interests which would be affected by the new act. After two years of negotiations, the minister of the environment reached a compromise with the industrial organisations and the associations representing regional and local authorities. However, he was not able to reach agreement with farmers (ibid., pp. 37-42). Farmer's associations supported the general aims of environmental policy, but rejected many of the specific measures in the new act. In particular, the associations could not accept that farmers were not entitled to compensation in cases in which they were forced to close down their farms for environmental reasons. Furthermore, the farmers argued that new appeal rules worsened their legal position. Therefore, they demanded compensation schemes and a better legal position in appeal cases. Since the government and the *Folketing* (the Danish parliament) would not meet these demands, farmers did not support the Law (Landbo-foreningerne, 1973, pp. 153-5; 1974, p. 108). In the *Folketing*, the majority was, nevertheless, careful not to hurt the interests of agriculture. It stated that the use of manure and chemical fertilisers could not be restricted unless there was provided sufficient knowledge on the environmental consequences of using manure and fertilisers. What is more, environmental demands on agriculture were not allowed to be unreasonable (Folketingets miljøudvalg, 1973, col. 2200).

Until the late 1970s, agriculture was, in most respects, exempted from pollution control (Andersen and Hansen, 1991, p. 48). To limit agri-pollution, the environmental authorities used 'soft' policy measures such as informative instruments and subsidies. The Environmental Protection Agency was careful to consult the farmers' associations on matters relating to agriculture. Usually, instructions to the local authorities and recommendations on the use of manure and fertilisers were the results of negotiations with farmers (Madsen, 1987, p. 74; Landboforeningerne, 1975, p. 122). Most confrontations between farmers and environmental authorities in the

1970s occurred in specific cases in which neighbours complained about the inconvenience caused by farms located in or near villages (Landboforeningerne, 1979, p. 95; Landbrugsraadet and Miljøministeriet, 1980, p. 13).

In the late 1970s, the character of the relationship between environmental authorities and farmers' associations changed from consultation and compromise to confrontation. In particular, ochre pollution of watercourses in Jutland triggered conflicts between the two parties (Andersen and Hansen, 1991, p. 52-4, Landboforeningerne, 1982, p. 125-6).[4] Nitrate pollution also became a major source of disagreement.

In the early 1980s, nitrate pollution could no longer be neglected. This was realised by the Ministry of the Environment which concluded that there was a great need of research into the causal links between fertiliser application and the quality of surface and ground water (Miljøministeriet, 1980). At the same time, the EPA drew attention to nitrate pollution from agricultural production. The establishment of an agricultural division resulted in a strengthening of the EPA's organisational resources in agricultural matters (Kampmann, 1980, p. 56-9). Farmers maintained that there was no pollution from agricultural production, claiming that environmental concerns were integrated into agricultural practices because farming was a human activity carried out in close contact with nature (Landboforeningerne, 1978, p. 95; 1980, p. 93). Farmers viewed themselves as custodians of the countryside whose way of life was in harmony with nature (interviews, Sørensen, the Smallholders' Union, 1993 and DFU, 1993a). Consequently, attempts to make farmers recognise that they did, in fact, pollute were met with massive opposition within the farming community. Later, they did, however, acknowledge that the use of manure and fertilisers could cause local pollution (Miljøministeriet and Landbrugsraadet, 1980, p. 22).

In the early 1980s, the relationship between farmers and environmental authorities was becoming more troublesome. The latter did not accept that agriculture was exempted from pollution control. Their efforts of the late 1970s and early 1980s aimed at ending agricultural exceptionalism in environmental politics but farmers were not willing to give up their special status.

First round: the NPO action plan

Strengthened research efforts within the EPA resulted in several reports on water pollution, published in the early 1980s. In 1984, the EPA presented the results of the research program in the NPO report ('N' stands for nitrogen, 'P' for phosphorus and 'O' for organic waste). It was produced by the EPA, but a steering committee consisting of civil servants from the EPA and the Ministry of Agriculture oversaw the preparation of the report.

Usually, Danish government reports try to give an impression of harmony and consensus, but the NPO report was unique in that it explicitly showed that there was a major conflict between the EPA and the Ministry of Agriculture. Basically, the two actors disagreed as to whether or not nitrogen application in farming should be defined as a political problem. The Ministry of Agriculture was convinced that it was not, whereas the EPA defined it as a serious political problem. As a result, they also disagreed on what policy measures should be implemented. The EPA estimated the nitrate run-offs from farming at 70-90 kilos per hectare, concluding that farmers caused considerable nitrate pollution (Miljøstyrelsen, 1984, p. 51). By contrast, the Ministry of Agriculture concluded that nitrate pollution was very limited, estimating nitrate run-offs to 53 kilos per hectare. Since non-farmed areas also leach nitrate, this natural run-off had to be deducted. Thus, the ministry concluded that the overfertilisation was only about 3 kilos per hectare (ibid., p. 45; Landbrugsministeriet, 1984). Not surprisingly, the farmers' associations agreed with these findings (Landboforeningerne, 1984, pp. 123-4). While the EPA claimed that agriculture was responsible for far the largest share of nitrate pollution,[5] the Ministry of Agriculture and the farmers argued that the EPA did not take into account that a large share of nitrate pollution could be put down to waste water from households (Landbrugsministeriet, 1986, pp. 18-19; interview, DFU, 1993a). The Ministry of Agriculture was prepared for conflicts with the EPA. Few years earlier it had set up a special unit to produce counter evidence to the EPA's research findings on agricultural pollution. Just before the EPA released the NPO report, the Ministry of Agriculture published its research results to the former's great surprise and annoyance (Pedersen and Geckler, 1987, p. 113).

To control nitrate run-offs from agriculture, the EPA suggested a legal requirement of 12 months' manure storage capacity, restrictions on farmyard manure applied per hectare, a fertiliser tax, and increased research and information on agri-environmental matters (Miljøstyrelsen, 1984, pp. 141-6). Environmental and financial considerations motivated the EPA to suggest the fertiliser tax. It believed that nitrate run-offs would decline if fertiliser prices increased because higher prices would force farmers to buy less chemical fertiliser and encourage them to use farmyard manure more effectively. Revenues from the fertiliser tax would be used to fund pollution control in agriculture (Miljøstyrelsen, 1985, p. 22). Most of these proposals were strongly opposed by the Ministry of Agriculture. In particular, it opposed the proposals on fertiliser taxes and on 12 months' manure storage capacity. Referring to its own investigations (Landbrugsministeriet, 1984), the Ministry of Agriculture argued that since there was no overconsumption of chemical fertilisers, there was no point in introducing a fertiliser tax which would merely be a tax on a production input with no significant environmental effect (Miljøstyrelsen, 1984, p. 8). Instead, the ministry recommended that politicians apply informative policy instruments which took into account the specific conditions of each farm (Landbrugsministeriet, 1986, pp. 5, 23, 32)

Officially, the Ministry of Agriculture participated in the steering committee overseeing the work on the NPO report to ensure that agricultural expertise was represented (Miljøstyrelsen, 1984, pp. 7-8). However, the real purpose was probably to make sure that the report took agricultural interests into account. This is clearly indicated by a statement from the vice director of the EPA who said that in environmental policy making, the Ministry of Agriculture was often more friendly to agriculture than was agriculture itself (Pedersen and Geckler, 1987, p. 112). Conflicting policy preferences led to a very bad relationship between the EPA and the Ministry of Agriculture with animosity towards EPA civil servants. Such feelings were explicitly expressed by the minister of agriculture who condemned the EPA's civil servants as advocates of special interests who acted as police authorities. Deeply distrustful of the motives of the EPA, civil servants in the Ministry of Agriculture believed agri-environmental policies should be handled by the ministry itself because it had the greatest agricultural knowledge (ibid., pp. 113-4; Christensen, 1987, p. 75; Landbrugsministeriet, 1986, p. 33). Neither had the farmers' associations any confidence in the EPA. The 1985 annual report of the Farmers' Union insinuated that the EPA's policy proposals were inspired by socialist ideals (Landboforeningerne, 1985, p. 107). As these were very unpopular among farmers, such attacks indicated the seriousness of the conflict. Using arguments similar to those of the Ministry of Agriculture, the Farmers' Union stated that a fertiliser tax was just a tax on raw materials used in agricultural production and as such would have no environmental effects (Landboforeningerne, 1986, pp. 107-8). Farmers believed that the primary purpose of fertiliser taxes was to fund the environmental bureaucracy (interview, DFU, 1993a).

The recommendations of the NPO report became the basis on which the EPA entered negotiations with the farmers' associations. Negotiators also included the Society for Nature Preservation, fishermen's associations and the Ministry of Agriculture. But as things turned out, the talks became a question of finding a compromise between the EPA and the farmers' associations (Madsen, 1987, p. 91, *Folketingstidende F*, 1984-85, col. 3881, 3891). They succeeded in this aim. The compromise clearly demonstrated that the EPA had given major concessions to the farmers. In the *Folketing*, the minister of the environment presented the compromise as the NPO action plan. It diverged considerably from what the EPA had originally recommended. The fertiliser tax was abandoned, the requirement on a minimum storage capacity was changed from 12 to five months, and the spreading of liquid manure was banned from harvest to the 15 October, unless the fields were covered with crops or about to be planted with winter crops. Restrictions on manure application were to be settled in negotiations between the Ministry of Agriculture and the agricultural associations (*Folketingstidende F*, 1984-85, col. 3879-3891). A compromise was possible because the EPA gave up on restrictions on manure and fertiliser application. These measures would limit nitrate run-offs from the fields. Thus, the controversial question on non-point pollution was downgraded and, consequently, the nitrate policy concentrated on point source pollution. In fact, the policy instruments which the two parties agreed on were, by and large, merely making the recommendations of the farmers' associations' advisory service binding. It had recommended six months storage capacity for manure and advised against spreading manure in September and October (Landbrugets Informationskontor, 1984, pp. 40-1)[6]

The majority of the *Folketing* did not accept the government's action plan. From 1982 to 1988, the government did not control a parliamentary majority in environmental politics. Hence, it was forced to enter a compromise with one of the parties belonging to the so-called green majority. By altering the plan, the government was able to meet some of the demands of that majority and could thus reach a broad agreement. The NPO action plan was, however, still based on regulatory and informative policy instruments. Storage of manure in the fields was to be prohibited; manure storage capacity was increased to six months; the rules for spreading of fluid manure were tightened so that manure spread on uncovered fields must be incorporated into the soil within 24 hours; restrictions on application of manure per hectare were specified and the need for research was specified. Despite broad parliamentary agreement on the measures, the battle on fertiliser taxes was not yet over. It was only possible to reach a compromise between the government and the opposition because the question of financing subsidies for the investments in storage facilities was left unsolved and postponed (*Folketingstidende C*, 1984-85, col. 717-20). The government supported a scheme in which the state funded 25 to 30 per cent of the expenses (*Folketingstidende F*, 1984-85, col. 3886), but the opposition supported a more stringent application of the polluter pays principle (Andersen and Hansen, 1991, p. 60).

It turned out to be difficult to settle the dispute. The minister of the environment had asked farmers' associations to draft a subsidy scheme for the increase of the manure storage capacity (Landboforeningerne, 1985, p. 110) but this strategy was not the way to build a parliamentary majority coalition. The *Folketing* spent nearly a year discussing the subsidy scheme before the government could achieve sufficient support for its bill. The green majority was in favour of a fertiliser tax as a means to fund agri-environmental subsidies. However, in the autumn of 1986, the government persuaded one party of the green majority, the Radical Liberal Party, to give up fertiliser taxes. The party had also been under heavy pressure from the Farmers' Union (Landboforeningerne, 1986, pp. 107-10). The subsidy scheme for investments in manure storage allowed farmers to choose between a state subsidy covering 25 to 40 per cent of the investments or an immediate writing off of the costs. The subsidy covered 40 per cent of the costs for small farms, 32.5 per cent for medium sized farms and 25 per cent for largest ones. The maximum was 100,000 *kroner* (Lov no. 16, 1987; Landboforeningerne, 1987, pp. 62-5). In the farming community, there was a split on the fertiliser tax. The president of the Smallholders' Union in an attempt to reach a political compromise suggested a low fertiliser tax, but had to withdraw the proposal because of opposition not only within the union (interview, Sørensen, the Smallholders' Union, 1993), but presumably also because of strong opposition from the Farmers' Union (Landboforeningerne, 1986, p. 107).

Danish farmers avoided the introduction of a high cost nitrate policy. They did not seem to feel that the policy was costly. As the former president of the Smallholders' Union said in an interview: 'we could live with the NPO action plan' (interview, Sørensen, the Smallholders' Union, 1993, my translation). The choice of policy instruments was not the only victory for the farmers. Since the NPO action plan did not specify any policy objectives, it was difficult, if not impossible, to evaluate.

Farmers used their power in the implementation process to lessen the effects of the policy instruments. The *Folketing's* strengthening of the government's original NPO action plan did not have the backing of the farmers' associations. Therefore, they used their power in the negotiations on the details of the plan to persuade the EPA to specify the policy measures in a way which was acceptable to them (Husmandsforeningerne 1985, pp. 43-4). More importantly, the farmers' associations reached an agreement with the National Association of Local Authorities which laid down guidelines for the local authorities' inspections of the farms and also established the criteria for choosing the officials who should inspect the farms. The agreement also recommended that each local authority and farm association set up a committee consisting of elected representatives to monitor the local implementation process (Landboforeningerne, 1985, pp. 110-11).

The Agricultural Commission

While the EPA was finishing the NPO report in the summer of 1984, the minister of agriculture set up the Agricultural Commission to investigate structural and environmental matters in Danish agriculture (Betænkning no. 1078, 1986, pp. 1-2). By doing this, the members of the agricultural policy network made a clear attempt to gain control over the agri-environmental policy process, realising that the pollution control could not be excluded from the political agenda any longer.

The Agricultural Commission set up two sub-committees, one for structural matters and the other for environmental matters. Both the commission and the sub-committee on environmental policy were strongly in favour of agriculture. Environmental interests were not represented in the commission at all, and in the sub-committee on environmental policy, only one third of representatives belonged to the environmental community (ibid.). The dominance of agricultural interests was clearly reflected in the commission's reports. Disagreements over the environmental sub-committee's conclusions and solutions were manifest. The EPA argued that a fertiliser tax was an effective instrument to limit nitrate run-offs, but the agricultural majority disagreed (ibid., pp. 101-3, 158-62). The environmental representatives made minority statements in which they expressed their dissatisfaction with the report's conclusions and recommendations (ibid., bilag 2 and 4).

The Agricultural Commission was not able to bring forward solutions to the pollution problems which could satisfy environmental interests. It argued that informative instruments such as strengthening the advisory service and more research could solve most of the problems. Regulatory instruments could be applied but they should first and foremost be seen as guidelines. While the commission refused to apply green taxes, it recommended the use of subsidies (ibid., in passim, in particular pp. 215-8). However, environmental pressure had increased too much to satisfy environmental interests by the use of informative instruments and state subsidies. Thus, the agricultural network was not able to co-opt the new issue area. Shortly after the Agricultural Commission's report was published, a bucketful of dead lobsters shocked agricultural policy makers.

In October 1986, fishermen caught some dead lobsters in Kattegat (the sea between Skagerrak and the Baltic). The lobsters had died of oxygen shortage which was ascribed to insufficient restriction on nitrogen application in agriculture. The subsequent months became a nightmare for the members of the agricultural policy network. They lost control of the situation and the policy process soon moved from the political-administrative arena to the cabinet and the *Folketing*. It turned out to be difficult to reach a compromise in these arenas. The policy process became highly politicised and occasionally chaotic. This was, to a large extent, the result of political parties' attempts to demonstrate their concern for the environment. Opinion polls in 1986 showed that voters were very attentive to the environmental issue, and an election was imminent (Andersen and Hansen, 1991, pp. 69-73).

The Danish media, particularly the TV-news, played an important role in putting nitrate pollution on to the political agenda. The director of the EPA's Marine Laboratory also had a central role in the agenda setting phase. His activity in the media was not, however, supported by the EPA which unsuccessfully tried to muzzle him. In the TV-news on Sunday the 19 October 1986, the Society for Nature Preservation was able to pressurise the government into taking action (Andersen and Hansen, 1991, pp. 73-84). The Society presented a six-point plan calling for a 50 per cent reduction in agricultural pollution within two years. This plan was drawn up in a telephone conversation the same Sunday morning as it was presented (Svold, 1989, pp. 73-4). Nine days later, the government presented an action plan which suggested a strengthening of inspections on point pollution. The government assessed that the use of chemical fertilisers could be reduced by 100,000 tons and suggested that specific measures to decrease fertiliser consumption should be negotiated with the farmers' associations (Miljøministeriet, 1986, pp. 6-8). The plan did not suggest green taxes or any other specific policy instruments to limit non-point pollution. The Farmers' Union criticised the government for suggesting specific policy objectives for the reduction in the consumption of chemical fertilisers (Landboforeningerne, 1987, p. 76). Nor did the action plan satisfy the green majority in the *Folketing*, but for different reasons: the plan was too lax. The green majority, therefore, forced the government to work out a new plan which could reduce nitrate run-offs in agriculture by 50 per cent within three years (Miljøministeriet, 1987, p. 11).

The cabinet set up an ad hoc cabinet committee to deal with the problem. Usually, the minister of the environment is appointed as chairman in committees dealing with pollution control but in this instance, the prime minister headed the committee. This reflected the serious conflict between the ministers of agriculture and of the environment (Akademiet for de Tekniske Videnskaber, 1990, pp. 41-2). A few months earlier, the Agricultural Commission had published the report in which the farmers' associations and the Ministry of Agriculture maintained that nitrate pollution was not a serious problem in farming. Therefore, the commission concluded that informative instruments were sufficient to persuade farmers to use nitrogen in an environmentally sustainable way (Betænkning no. 1078, 1986, pp. 22, 215-8).[7] The

EPA viewed agricultural nitrate pollution as a serious problem (Miljøstyrelsen, 1984, p. 51). These conflicting views were represented in cabinet by respectively the minister of agriculture and the minister of the environment (Akademiet for de Tekniske Videnskaber, 1990, p. 41; Christensen, 1987, p. 75).

In early December 1986, the government also appointed a committee of experts to prepare a revised action plan. This consisted of representatives from the EPA, the farmers' associations, the Ministry of Agriculture and the Council of Science. Although the Society for Nature Preservation had played a major role in the agenda-setting phase, it was not invited to participate. The task of the committee was to examine how the *Folketing*'s decision could be put into operation. The committee concluded that it would be unrealistic to reduce nitrate run-offs in agriculture by 50 per cent within three years. Because the committee's members disagreed on the extent of over-fertilisation, it could not reach agreement on the choice of policy instruments. The EPA supported fertiliser taxes and quotas while the Ministry of Agriculture and the Farmers' Union opposed the use of such measures. The Smallholders' Union did not explicitly oppose taxes and quotas, but with the Farmers' Union it criticised the way in which the preparation of the report was conducted (Miljøstyrelsen, 1986). Both the farmers' and the smallholders' unions preferred informative instruments and environmental state subsidies to control nitrate run-offs (Landboforeningerne, 1987a, p. 2; 1987b; Husmandsforeningerne, 1987). These measures were in line with those already recommended by the Agricultural Commission.

The inability of the expert committee to reach agreement on a revised action plan meant that the decision making process moved to the cabinet. However, organised interests were still in close contact with the ministries (Akademiet for de Tekniske Videnskaber, 1990, p. 44). On 31 January 1987, the government presented a revised action plan to the *Folketing* and then a somewhat chaotic period followed. The Farmers' Union leaked a confidential report which threw doubt on the data underpinning the plan. Clearly, the aim was to have the plan re-negotiated. However, this move had the opposite effect. It fuelled the conflict between the minister of agriculture and the minister of the environment. The latter, along with his party, threatened to leave the government coalition.[8] In the *Folketing*, the attempt to throw doubt on the action plan influenced the green majority to reject the new action plan. It demanded that the plan be strengthened by a threat of a fertiliser tax if farmers did not decrease the consumption of chemical fertilisers within three years. Dissatisfied with the plan, the green majority also decided that the *Folketing's* Environment and Planning Committee should take over the work on the action plan (Andersen and Hansen, 1991, pp. 102-9, Akademiet for de Tekniske Videnskaber, 1990, pp. 45-7).

After long, complicated discussions in the Environment and Planning Committee, the *Folketing* approved the aquatic action plan in June 1987. Considerable pressure from smallholders and farmers within the Radical Liberal Party forced the party's environmental spokes-person to give up the demands for fertiliser taxes (*Press*, no. 37, 1988). Smallholder interests have traditionally been represented by the Radical Liberal Party. Giving up the fertiliser tax meant that the government could reach a compromise with the radical liberals. The final aquatic action plan applied mostly regulatory policy instruments. The most important of these were: the legal requirement on manure storage capacity was increased from six to nine months (this requirement could under certain conditions be adjusted to individual conditions), a legal requirement on certain percentages of autumn crop cover (45 per cent in 1988, 55 per cent in 1989, 65 per cent in 1990), a legal requirement for the farmer to work out a fertiliser and crop rotation plan, and a legal requirement to incorporate fluid manure into the soil within 12 hours. The period in which the environmental subsidy scheme for the enlargement of manure storage facilities was in force was extended (Miljø- og planlægningsudvalget, 1987). In addition, farmers were promised subsidies for planting catch crops;[9] however, this scheme was never implemented (De danske Landboforeninger and Danske Husmandsforeninger, 1990, p. 14).

Farmers had avoided constraining policy objectives in the NPO action plan. They were less successful in the aquatic action plan; however, they could have been worse off had the *Folketing* enacted specific objectives for the reduction of fertiliser consumption (included in the government's first draft for an action plan). Farmers were more amenable to the policy objective stating that nitrate run-offs should be reduced by 50 per cent within three years (Andersen and Hansen, 1991, p. 92). However, they did not believe that this could be achieved within three years and, consequently, demanded a longer period to achieve it (Landboforeningerne, 1987b; Husmandsforeningerne, 1987). The Farmers' Union said that run-offs could at best be reduced by about 40 per cent. The reason farmers preferred the run-off aim to the consumption goal was that the main participants in agri-environmental policy making, even the EPA, regarded the run-off objective as unrealistic. Moreover, it would be extremely difficult to measure whether or not the objective was fulfilled. Thus, farmers had every reason to believe that it would have no practical consequences.

Originally, the government's intention was that the Ministry of Agriculture should specify the requirements on autumn crop cover (Miljøministeriet, 1987, p. 17), but the green majority could not accept this. To reach agreement on policy measures in the *Folketing*, the aquatic action plan left that problem unresolved. However, the green majority could not prevent the prime minister from using his constitutional right to give this problem area to the Ministry of Agriculture (Akademiet for de Tekniske Videnskaber, 1990, pp. 57-8). This was a major victory for the members of the agricultural policy network because they gained control over a large part of the implementation of the aquatic action plan. The use of policy instruments could then be influenced by agricultural concerns.

The Ministry of the Environment retained the responsibility for revising the departmental order regulating manure storage facilities. The farmers' associations did not accept the requirement on nine months' storage capacity. After almost a year of discussions with the minister of the environment, there was still no solution to the problems, but after the general election in 1988, the situation changed. The prime minister succeeded in splitting the green majority in the *Folketing* by forming a government coalition with one of the constituent parties from the green majority. As a result, the farmers' associations were able to convince the new minister of the environment, who was a radical liberal, that the requirement on manure storage capacity should be more flexible. The Minister accepted that local government would automatically allow manure storage capacity for less than nine months if the agricultural advisory service recommended it. To achieve such permission, the fertiliser and crop management plans should state that manure could be applied in accordance with the aquatic action plan's rules on manure application (Bekendtgørelse no. 568, 1988; Landboforeningerne, 1988, p. 69). This change in the rules gave farmers considerable influence in the implementation process. The advisory service - which the farmers' associations run - was the only organisation possessing the bureaucratic resources to assess the need for storage capacity on about 30,000 livestock farms (Dubgaard, 1991, p. 36). Thus, public authorities had to rely on the information provided by the advisory service.

Farmers also persuaded the minister of the environment to administer the specific rules on incorporation of manure into the soil in a less stringent way. To compensate farmers for the costs of the nitrate policy, the minister of agriculture provided indirect environmental subsidies to farmers, for example increased economic support for the advisory service and reductions in farmers' payments for veterinary inspection (Landboforeningerne, 1988, pp. 70-1, Husmandsforeningerne, 1988, pp. 38-9). Furthermore, the environment and agriculture ministers agreed not to enforce the legal requirement on fertiliser and crop rotation plans (Landbrugsministeriet, 1991, p. 278). In other words, they altered the measure from a regulatory to a voluntary policy instrument.

The farmers' associations used their political power in the implementation process to obtain environmental subsidies, changing universal regulatory instruments to individualised instruments. Moreover, they managed to change a regulatory instrument into an informative instrument. In other words, farmers succeeded in diminishing their political and economic costs involved in the implementation of the nitrate policy. A high degree of politicisation in the policy formulation process resulted in policy choices unacceptable to farmers but the depoliticised implementation process, so to speak, brought the system back into its own 'equilibrium'. The aquatic action plan required the government to present a report on the implementation of the plan in 1990. This account made it obvious that more measures needed to be taken if the ambitious objective of the plan (50 per cent reduction in nitrate pollution) was to be achieved.

Third round: the action plan for sustainable agriculture

The minister of the environment's report on the aquatic action plan in the spring of 1990 stated that nitrate pollution had only been reduced by 20 per cent instead of 50 per cent which was the objective.[10] Furthermore, the Minister also admitted that the policy instruments were insufficient to reach the aim. Consequently, the government would draft a new plan (*Folketingstidende F*, 1989-90, col. 9080-91).

The Ministry of Agriculture was given responsibility for producing a report on sustainable agriculture and drafting yet another action plan. This was a major victory for the members of the agricultural policy network, particularly for the Ministry of Agriculture which, after years of bureaucratic struggles with the Ministry of the Environment and the EPA, gained responsibility for a great deal of agri-environmental policy making.[11] Moreover, the administration of the environmental subsidy scheme was also handed over to the Ministry of Agriculture (Husmands-foreningerne, 1990, p. 24, Landboforeningerne, 1990, p. 34).

The Ministry of Agriculture did not, as one might expect, involve the farmers' associations in the preparation of the report, neither was the Ministry of the Environment and the EPA invited to take part in the work. These actors did, however, have the opportunity to comment on the report before it was published (Husmandsforeningerne, 1990, p. 26; interviews, Ministry of Agriculture 1993, Ministry of the Environment, 1993 and DFU, 1993b). The reason for not consulting the agricultural organisations might have been the experiences of the Agricultural Commission four years earlier. It did not come up with results convincing enough to allow the Ministry of Agriculture to gain control over the nitrate policy. The experience of the Agricultural Commission demonstrated that agri-environmental policy making could not just be carried out as traditional agricultural policy making in which a compromise between the ministry and the agricultural associations was essential for the contents of policy. Environmental concerns had to be upgraded as the issue was still highly politicised. The ministry had to respond convincingly to the problems of non-point nitrate pollution if it wanted to stay in charge of policy making within the issue area.

While the farmers' associations, the Ministry of the Environment and the EPA only had a very limited say on the contents of the report on sustainable agriculture, the Ministry of Agriculture was more willing to listen to them when it drafted the action plan for sustainable agriculture (interviews, Ministry of Agriculture 1993, Ministry of the Environment, 1993 and DFU, 1993b). Nevertheless, the plan clearly demonstrated that the Ministry of Agriculture had the final word. The Society for Nature Preservation, which had played an important role in the agenda setting phase of the earlier aquatic action plan, did not have much influence on the substance of the nitrate policy. However, the Ministry of Agriculture briefed the Society several times about the work on the report (interview, Ministry of Agriculture, 1993). Thus, the Society was not totally excluded from the policy process, but from influence.[12]

Basically, the plan recommended that existing policy measures be tightened and suggested that subsidies should be used to extensify farming in environmentally sensitive areas. Informative instruments should be changed into regulatory instruments and existing regulatory instruments should regulate the behaviour of farmers more strictly. The Ministry of Agriculture recommended that each farmer should be legally obliged to make an annual fertiliser and crop rotation plan which balanced the supply of manure and chemical fertilisers with the demands of the crops. The main aim of these plans was to avoid over-consumption of fertilisers. To make sure that farmers complied, they should be legally required to render a fertiliser account. The Ministry of Agriculture also recommended that the rules on application of manure were tightened. To protect the ground water in areas in the regions of North Jutland, Viborg and Aarhus, the ministry suggested that the most environmentally sensitive areas were identified and farmers offered a subsidy to reduce the use of fertilisers and manure.[13] Finally, the ministry recommended that the period in which the subsidy scheme for the enlargement of manure storage facilities was in force was extended (Landbrugsministeriet, 1991, pp. 338-41).

These recommendations were, to a large extent, in agreement with the policy preferences of the farmers' associations. In their proposals for sustainable development in agriculture they suggested increased use of fertiliser and crop rotation plans as the major instruments to decrease nitrate run-offs. They could not accept fertiliser taxes, arguing that they were bureaucratic, had no environmental effects and that they put unnecessary economic strains on farmers. Farmers regarded a fertiliser tax as one with mainly fiscal purposes (Landboforeningerne, 1991a, pp. 4, 11; Husmandsforeningerne, 1991, pp. 22-5; De danske Landboforeninger and Danske Husmandsforeninger, 1990, pp. 6, 24). While the Ministry of Agriculture argued that the preparation of fertiliser and crop rotation plans and accounts should be obligatory, the farmers' associations preferred these measures to be voluntary. Thus, though the ministry went a little further than farmers believed was necessary (interview, DFU, 1993b), in practice the measures were almost identical.

To meet demands from the EPA, the report on sustainable agriculture analysed the economic and environmental effects of fertiliser taxes and quotas. It concluded that both policy instruments had negative effects on the distribution of incomes in agriculture and also on the choice of crops. In order to avoid these negative effects, the administration of the instruments would be relatively complicated (Landbrugsministeriet, 1991, pp. 228-42). Therefore, the Ministry of Agriculture recommended the use of regulatory, informative and positive economic policy instruments. The EPA and the Ministry of the Environment had previously argued that fertiliser taxes should be introduced. In the preparation of the action plan for sustainable agriculture, they kept a much lower profile than in the policy processes of the two previous action plans. However, they did, and still do, prefer negative economic instruments, such as fertiliser taxes (interviews, Ministry of the Environment, 1993 and the EPA, 1993). The Society for Nature Preservation chose a hard line, arguing that fertiliser taxes, quotas and strict rules on manure application were necessary to limit nitrate run-off significantly (Danmarks Naturfredningsforening, 1991, pp. 84-8). This stance may be the reason the Society could not influence the substance of the nitrate policy.

In 1990, the government coalition changed. The radical liberals left the government after a serious defeat in the 1990 general election and then the government consisted only of liberals and conservatives - a weak minority government. Therefore, the *Folketing* could influence the measures of the action plan for sustainable agriculture. Unlike the negotiations on the two previous action plans, the *Folketing* succeeded in handling the situation without major conflicts. In 1991, a special committee, chaired by the former social democratic minister of agriculture, was set up to deal with the action plan in 1991. The committee was composed of members of

96

parliament from the Environment and Planning Committee and from the Agricultural and Fisheries Committee. It accepted most of the measures of the government's action plan but not the rules on application of farmyard manure. Consequently, the committee tightened these rules by setting targets for the exploitation of nitrogen in manure (Udvalget vedrørende en Bæredygtig Landbrugsudvikling, 1991). The following year, the committee accepted that the government individualised the rules on manure storage capacity with the effect that the fertiliser and crop rotation plan became the basis of the assessment of storage capacity. Officially, the minimum storage capacity was six months (Bekendtgørelse no. 1121, 1992) but, unofficially, the storage capacity could be only four months (interviews, Ministry of the Environment, 1993 and DFU, 1993b). Farmers had asked for this alteration (Landboforeningerne, 1991, p. 44). The *Folketing* decided that the period in which the subsidy scheme for the increase in manure storage capacity was in force was prolonged until 1995 and that the maximum payment be increased from 100,000 to 150,000 *kroner*. In addition, the regulations were changed so that all farms were entitled to a 35 per cent subsidy (Lov no. 1172, 1992).

When the minister of the environment gave an account on the aquatic action plan in the *Folketing* in 1990, she said that its objectives had not been achieved and new policy measures were needed (*Folketingstidende F*, 1989-90, col. 9080-91). As early as 1987, most policy makers had recognised that the objectives were unrealistic. As a consequence, the deadline for the achievement of the 50 per cent reduction of nitrate run-offs was postponed for ten years, that is, to the turn of the century (Landbrugsministeriet 1991, p. 343).

Sweden[14]

Sweden was one of the first countries in the world to have an environmental protection law. When the Environmental Protection Law was adopted in 1969, the Swedes were four years ahead of Denmark. In contrast to the Danish Environmental Protection Law, the Swedish did not exempt farmers from pollution control. Farmers had positive attitudes to environmental protection and, therefore, they had no difficulties accepting the new law.

It drew attention to agricultural point source pollution by stating explicitly that it was illegal to discharge animal urine into surface waters. The departmental order accompanying the Environmental Protection Law required farms with more than 1,000 pigs or 10,000 poultry to apply for a farming licence (SFS 1969: 387; SFS 1969: 388). Before this was issued, the County Administrative Board (*Länsstyrelsen*) could require that the applicant had sufficient acreage for manure spreading, manure storage capacity for five to six months and that manure be incorporated into the soil within 24 hours (Naturvårdsverket, 1970, pp. 6-8). However, 15 years after the adoption of the Environmental Protection Law, the requirement on manure storage capacity had not yet been implemented on all farms (Proposition 1987/88: 128, p. 25).

The farmers' associations supported the Environmental Protection Law (Proposition 1969: 28) and participated in the preparation of its implementation guidelines (Naturvårdsverket, 1970, p. 3). Until 1977, the state refunded 25 per cent of the costs of environmental investments in the manufacturing sector. Farmers were not entitled to environmental subsidies. They argued that they also had the right to such support. In 1972, they persuaded politicians to extend the subsidy scheme to the agricultural sector (Proposition 1972: 79, pp. 19-21, 29; SFS 1972: 293). However, farmers seem to have paid a certain political price for this achievement. It seems that the price was that farms with more than 100 livestock units should apply for an environmental licence (SFS 1972: 224). This was a tightening of the nitrate policy.[15]

In the beginning of the 1970s, Swedish farmers were much more positive towards environmental policy than their Danish counterparts. They introduced their own environmental programme in 1972 (LRF, 1972). This programme and other statements from the late 1960s and early 1970s demonstrated why farmers did not attempt to exclude the issue of pollution control from the political agenda. Their view on pollution did not conflict with their own production methods as the vice president of the Federation of Farmers, LRF, pointed out in 1972 by saying: 'farming and the environment are in fact one and the same' (*Land* 1972, no. 29, my translation). Farmers did not believe that they were causing pollution; industry and car owners were the real polluters (interview, de Woul of LRF, 1994). While farmers chose an offensive strategy in environmental policy making, the National Board of Agriculture was sceptical of pollution control in agriculture. It perceived environmental concerns as a barrier to its main objective which was the rationalisation of agricultural production. The Board, therefore, tended to focus solely on the economic efficiency of agricultural production (interviews, EPA 1994 and Board of Agriculture, 1994).

Farmers chose an offensive environmental strategy for political rather than for environmental reasons.[16] Consumer interests and market concerns were becoming more important politically. The social democrats, who were in power from the early 1930s until 1976, had disassociated themselves from agricultural interests in the 1960s (Micheletti, 1990, pp. 87-101; Elvander, 1969, pp. 243-4; Steen, 1988, pp. 310-3). Farmers felt that they were losing their traditional supporters (interview, de Woul of LRF, 1994) and, thus, had to mobilise the support of new groups. At that time, the environmental issue seemed to fit this objective very well; the issue was gaining public attention and it did not seem to hurt farmers themselves as they believed that modern farming was in harmony with nature.

The farmers' offensive strategy had some risks. Agricultural pollution could be difficult to oppose. For instance, questioning the use of chemicals could cause political difficulties for farmers (see Micheletti, 1990, pp. 68-9). In 1972, such a difficulty occurred when the minister of agriculture set up a commission to investigate the use of chemicals (pesticides and fertilisers) in agriculture. Two years later, the commission published its report, concluding that the use of fertilisers constituted hardly any threat to the natural environment. The commission also concluded that a prohibition on the use of fertilisers would have serious consequences for Swedish society. General restrictions would require a re-assessment of the agricultural policy. To meet

98

future demands for regulation of fertiliser use, the commission recommended more research into the environmental effects of fertiliser application and more advice to farmers (SOU 1974: 35, pp. 11-16). The report did not, however, lead to further political action.

During the second half of the 1970s, politicians in the *Riksdag* (the Swedish parliament) occasionally discussed the use of fertilisers. In particular, they considered nitrate in drinking water a serious problem and urged public authorities to monitor it very closely (JoU 1976/77: 26, pp. 25-6). In 1978, the social democrats made a statement in the *Riksdag* on the risks caused by nitrate polluted drinking water and asked for a commission to be set up to investigate nitrate pollution. They also listed effective policy instruments which could be applied, for instance a fertiliser tax (JoU 1978/79: 16, pp. 6-8). Discussions on fertiliser taxes in the *Riksdag* showed that farmers were beginning to lose control over agri-environmental policy making. Green taxes are the last thing farmers want. The offensive strategy which had earlier seemed to be without risks had turned out to be risky. In 1979, farmers again came under pressure when the minister of agriculture set up a commission to investigate the environmental, economic and employment effects of a reduced application of chemicals and fertilisers (SOU 1983: 10, p. 265).

First round: the defeat of the farmers

The commission was named the Commission on Application of Chemicals in Agriculture and Forestry. It finished its work in 1983, concluding that the use of fertilisers in agricultural production caused pollution problems in both ground and surface water. Furthermore, the commission estimated that more than 100,000 people were drinking water containing too much nitrate. To limit nitrate pollution, the commission suggested a number of policy measures. The majority in the commission argued that fertiliser taxes and better advice to farmers were suitable policy measures to limit the use of chemical fertilisers and to motivate farmers to apply farmyard manure more efficiently. Moreover, the majority suggested that advice and inspections should be concentrated in environmentally sensitive areas. A new law should authorise the government to point out such areas and the county administrative boards, which reported to the Environmental Protection Agency (EPA), would announce the measures to be adopted. The revenue from the fertiliser tax would fund pollution control in the agricultural sector (SOU 1983: 10, pp. 19, 233-52). Surprisingly, the National Board of Agriculture, which had been very sceptical of pollution control in the 1970s, formed part of the commission's majority and, thus, left farmers isolated in the policy process. The reason the Board did not oppose fertiliser taxes might have been that it would benefit from the tax revenues. To improve the advisory service, the commission suggested that these revenues should fund a staff increase in the county agricultural boards (*Lantbruknämnderna*) (ibid.: 236-8, 251-2). These boards gave advice to farmers and were the Agricultural Board's regional implementation bodies. Since 1967, the Swedish state has run most of the agricultural advisory service through the county agricultural boards (SOU 1992: 99, pp. 20, 58; see also Rothstein, 1992, pp. 241-6).[17]

Not surprisingly, the commission's farmer representatives opposed the majority's proposals. Their minority statement pointed out that the report lacked sufficient documentation, that it did not account for pollution from manufacturing and from the households at all and that, in general, the report 'went over the top' (my translation). They opposed fertiliser taxes and argued that more advice to farmers was the most effective policy instrument to cope with nitrate pollution. Furthermore, the Farmers' Federation did not want the county administrative boards (the regional implementation body of the EPA) to implement nitrate policy measures in agriculture. Instead, it wanted the Agricultural Board and its regional implementation bodies, the county agricultural boards, to do this. Finally, the farmer representatives wanted more research (SOU 1983: 10, pp. 297-305). The report clearly showed that pollution control had turned into a political problem for farmers.

The report of the commission laid down the foundations of the nitrate policy making for the next two years. Basing its bill on the commission's report, the social democratic government proposed a 5 per cent fertiliser tax in 1984 (Proposition 1983/84: 176). In the *Riksdag*, the Centre Party and the Conservative Party opposed such a tax (JoU, 1983/84: 15, bilaga 7, pp. 25-6), but this opposition was not sufficient to defeat the bill. Farmers might have been in a weak position because a fertiliser tax was introduced in 1982 to fund export subsidies for surplus cereals. Hence, the fertiliser tax could easily be collected through the administrative system already established. Introducing the fertiliser tax seems to have been a symbolic act. Three years after its introduction, a state secretary in the Department of Agriculture admitted that nobody had considered to what the tax revenue should be assigned (*Land* no. 7, 1987). After some years of political controversy, the Federation of Farmers succeeded in having part of the revenue spent on agri-environmental research (LRF, 1990, p. 31).

Later in 1984, farmers were overruled again. The Commission on the Application of Chemicals in Agriculture and Forestry had recommended that the government point out environmentally sensitive areas in which special measures should be put into effect. This recommendation became the official policy of the government. The government bill, which was largely the commission's proposal, had the support of the EPA and the Agricultural Board. Farmers opposed it, arguing that existing measures were sufficient because the requirements on storage capacity for farmyard manure had earlier been tightened (Proposition 1984/85: 10, pp. 35, 42). Apparently, the reason why the Agricultural Board supported the bill was that it would gain responsibility for giving advice to farmers in the environmentally sensitive areas. To carry out this task, the Board would receive more bureaucratic resources (SOU 1983: 10, pp. 236-8). What is more, it would achieve a certain control over the implementation of the nitrate policy. The specific role of the Agricultural Board was, in cooperation with the EPA, to work out guidelines for local implementation. However, the Board did not succeed in gaining full control over the nitrate policy in the environmentally sensitive areas, since the power to specify policy measures remained with the county administrative boards. The policy instruments which they could apply were all regulatory; for instance, compulsory manure and fertiliser planning, compulsory requirements on storage capacity for farmyard manure and specifications for the size of the area required for the spreading of manure (Proposition 1984/85: 10, pp. 12-13, 16). Farmers would receive no economic compensation for the costs of implementing the policy measures (LRF, 1985, p. 12; *Land* no. 45 1987). The Swedish nitrate policy of the early 1980s did not set any policy objectives. This was an advantage for farmers since it meant that their environmental achievements would be difficult to evaluate in the future.

By the end of 1984, environmental regulation in agriculture had changed. From being only a matter of point source pollution on large farms, the nitrate policy now aimed at non-point pollution. Farmers had to accept the introduction of policy instruments which aimed at limiting the use of farmyard manure and chemical fertilisers. In the early 1970s, policy makers discussed the issue but found no reason to legislate. In the late 1970s and early 1980s, farmers began to lose control over the agenda. From being able to influence discussions on pollution control in a way which directed attention away from agricultural pollution, farmers came to see that environmentalists, in particular the EPA, gradually achieved control over the agenda.

Second round: extending the nitrate policy and reaching a compromise

In 1986, nitrate pollution again appeared on the political agenda. The social democratic government decided to set up the Task Force against Pollution of Marine Waters (*Aktionsgruppen mot havsföroreningar*) consisting of six ministers (including the minister of agriculture), top civil servants from the agencies concerned (including the EPA and the Agricultural Board) and scientists to investigate the pollution of marine waters. Its task was to prepare an action plan. Birgitta Dahl, who later became minister of the environment, was appointed chairman (Naturvårdsverket, 1987: 2, 73). In May 1987, the final report was presented. In agriculture, the aim was to reduce nitrate run-offs by 50 per cent by the year 2000; however, the policy instruments suggested would only bring about a 20 per cent reduction; the remaining 30 per cent was supposed to come from changes in the agricultural policy and from other unspecified policy measures. Regulatory policy instruments such as rules for manure application and for manure storage capacity, increased use of catch crops, authorisation of manure spreaders and fertiliser distributors, and increased use of crop and fertiliser rotation plans were the core of the report's recommendations (Naturvårdsverket, 1987: 52-4). Both agricultural and environmental authorities represented in the task force agreed on these policy instruments.

The Department of Agriculture also addressed agri-environmental policy issues. In spring 1987, the minister of agriculture set up a committee, named the Intensity Group, to suggest changes in the agricultural policy. It examined how surplus cereals production, pollution and environmental damage in the agricultural sector could be decreased. The Federation of Farmers, the Agricultural Board, the EPA, the Agricultural Marketing Board, the Swedish Society for Nature Protection, the Department of Agriculture, consumer representatives and agricultural scientists were all represented in the committee (Jordbruksdepartementet, 1987). The majority of the Intensity Group concluded that by lowering the intensity of cereals growing, both surplus production and environmental problems would decrease. To cope with the latter, the committee recommended that the government increase the existing fertiliser tax and pay farmers to grow catch crops. It also recommended the use of regulatory policy instruments in environmentally sensitive areas and of more informative instruments (Ibid.: 68-75). Compared to the situation in Denmark, where the Ministry of Agriculture was overtly sympathetic to the interests of farmers, it is remarkable that Swedish agricultural authorities did not oppose fertiliser taxes. They accepted fertiliser taxes as long as they were funding pollution control in agriculture (Proposition 1987/88: 128, bilaga 2, pp. 85-6). However, they argued that the environmental effects of fertiliser taxes were very modest (interview, Board of Agriculture, 1993; see also Jordbruksverket, 1992). To give farmers better environmental advice, the Intensity Group recommended that the Agricultural Board's budget be raised by 5 millions *kronor* (Jordbruksdepartementet, 1987, p. 74). This was a major reason for the Agricultural Board's support for the recommendations of the report. Not only would its budget increase, it would also increase its number of employees since it ran the agricultural advisory service.

The farmer representatives belonged to the minority in the Intensity Group. They could not accept increases in the fertiliser tax. Instead, they recommended that the nitrate policy apply regulatory policy instruments which considered regional and local conditions (Jordbruksdepartementet, 1987, pp. 83-6). Farmers criticised the report for focusing too much on chemical fertilisers. They wanted policy makers to pay more attention to farmyard manure which they considered the most important source of nitrate run-offs. In addition, they demanded compensation for the costs of implementing new policy measures (Proposition 1987/88: 128, bilaga 2, pp. 67, 82). The Swedish Society for Nature Protection also made a minority statement in which it demanded that the government increase the fertiliser tax considerably and that it subsidise farmers who extensified their production (Jordbruksdepartementet, 1987, pp. 83-4).[18] However, since the Society lacked technical resources, it had only a very limited influence on the nitrate policy (interviews, Swedish Society for Nature Protection, 1993, EPA, 1993, 1994, Board of Agriculture, 1993, 1994, Department of Agriculture, 1993 and Department of the Environment and Natural Resources, 1993). Similarly, consumer representatives did not exercise much influence on the nitrate policy (interviews, EPA 1994 and Board of Agriculture, 1994).

The reports of the Task Force against Pollution of Marine Waters and of the Intensity Group became the foundation of the social democratic government bill on agri-environmental policy in the early spring of 1988. The government had ensured that a majority in the *Riksdag* supported it before it was introduced. The bill, to some extent, took farmers' interests into account but they did not get all they wanted. The Federation of Swedish Farmers had prepared its own action plan for reducing nitrate run-offs. It recommended the employment of individualised regulatory instruments, the core being compulsory crop and fertiliser rotation plans. These plans should also be the basis for assessing the need of manure storage capacity on each farm. Farmers should be entitled to subsidies covering 50 per cent of the costs of enlarging manure storage facilities if they made the investment before 1992. Farmers who increased their manure storage facilities after the end of 1991 should only be entitled to subsidies covering 25 per cent of the costs. Although instruments taking individual conditions into account were complicated solutions, the Federation of Farmers preferred them to more universal ones (Proposition 1987/88: 85, bilaga 5.2, pp. 137-8; bilaga 10.3, p. 443; *Land* no. 25, 1987).

Farmers did not succeed in having it this way; instead, they had to accept that the government adopted regulatory policy instruments which took only regional conditions into account. As a result, in some regions, the legal requirement for manure storage capacity would be eight to ten months for farms with more than ten livestock units and six months for farms with less than ten livestock units. In other regions, the requirement on eight to ten months storage capacity was only in force for farmers with more than 100 livestock units; farms with fewer livestock units were only required to have storage capacity for six months. In the most environmentally sensitive areas, the government recommended that crops covered 60 per cent of the farm land in the autumn. Farmers could not persuade the government to grant subsidies covering 50 per cent of the costs of enlarging manure storage facilities. The government decided to pay only 20 per cent of the costs, the maximum being 25,000 *kronor*. The subsidy scheme would only be in force from 1989 to 1991. While farmers may have perceived their inability to persuade the government to adopt individualised regulatory instruments and high subsidies as a political defeat, doubling the fertiliser tax was an even greater defeat. To decrease the use of fertilisers and stimulate more effective use of farmyard manure, the government recommended that the *Riksdag* increase the fertiliser tax to 10 per cent. The purpose of the tax was also to fund environmental policy measures in agriculture, including agri-environmental subsidies. The bill also contained some victories for farmers. They succeeded in persuading the government to provide farmers with advice on the preparation of fertiliser and crop management plans and also to pay subsidies for the sowing of catch crops. Farmers had until 1995 to meet the requirements of the nitrate policy (Proposition 1987/88: 128, pp. 19-34; JoU 1987/88: 24; SFS 1989: 12). LRF, the Farmers' Federation, seemed to be content with the choice of policy instruments.[19] In particular, they emphasised that they had avoided universal instruments and that the revenue of the increased fertiliser taxes would be refunded to farmers through agri-environmental subsidies (*Land* no. 9 1988; Proposition 1987/88: 128, p. 40). However, farmers still opposed the use of fertiliser taxes (*Land* no. 12 1988).

The policy process focused on the choice of instruments rather than on the choice of policy objectives. Unlike the first round of agri-environmental policy making, the policy process in the second round produced a set of policy objectives. Firstly, the government and the *Riksdag* accepted the overall objective proposed in the report of the Task Force Against Marine Pollution. Nitrate run-offs should be reduced by 50 per cent by the year 2000. Secondly, the nitrate policy operated with a specific objective: a reduction in the use of fertilisers. The *Riksdag* accepted the government's long term aim which was a 20 per cent reduction by the year 2000. The medium term goal was a 10 per cent reduction by 1992 (Proposition 1987/88: 128 pp. 11; JoU 1987/88: 24). Farmers accepted the overall objective of a 50 per cent reduction in nitrate run-offs and did not seem to oppose the other policy objectives (Proposition 1987/88: 85, bilaga 5.2, p. 126).[20]

The organisation of the implementation process ensured that agricultural concerns could influence the practical use of the policy instruments. Thus, to some extent, farmers could lessen the costs of the nitrate policy. The regulatory policy instruments were written into the Law on Protection of Farm Land (*Lag om skötsel av jordbruksmark*) (SFS 1988: 640). This meant that the Agricultural Board and the county agricultural boards would implement important parts of the nitrate policy. However, the specific rules were to be prepared in cooperation with the EPA. Agricultural interests would have faced a more disadvantageous situation had earlier organisational recommendations been carried out. A commission examining the organisation of environmental authorities had recommended that the rules of the nitrate policy be written into the Environmental Protection Law (SOU, 1987: 32, p. 337) with the effect that the EPA, the county administrative boards and the municipalities' health committees would have become the implementing authorities. But the minister of agriculture persuaded the minister of the environment to approve that the county agricultural boards become the implementation body (Proposition 1987/88: 128, pp. 40-1). Although the organisation of the implementation process favoured agricultural interests, the EPA and the county administrative boards still played an important role. They were responsible for issuing environmental licences for farms with more than 100 livestock units, and they had the authority to specify the policy measures in environmentally sensitive areas.

Since the mid-1980s, the Federation of Farmers has developed more positive attitudes to pollution control. For instance, in 1985, it recognised that nitrate run-off was a problem which farmers had to solve (LRF, 1985, p. 4). Later, the vice president of the Federation of Farmers said that farmers had to diminish nitrate run-offs and that they were striving to limit nitrate run-offs by 50 per cent by 1995 (Jönson, 1989, pp. 33-4). This deadline was five years ahead of the official one. In addition, farmers have launched a campaign aiming at making Swedish agriculture the cleanest in the world (LRF, 1992, p. 5). The positive environmental attitude of farmers is a major reason for the decrease in the conflict between environmental and agricultural interests (interviews, EPA 1994, de Woul of LRF, 1994 and Board of Agriculture, 1994). The cooperative strategy which the Federation of Farmers pursues demonstrates that Swedish farmers have realised that their political power is insufficient to achieve their objectives using confrontational strategies. They seem to believe that they will do better by pursuing compromise.

Third round: changing the nitrate policy incrementally

During the late 1980s, the Swedish environmental administration changed. To strengthen environmental policy making, the social democratic government estab-

104

lished the Department of the Environment and Energy and appointed a minister of the environment in 1987. Previously, the Department of Agriculture was assigned responsibility for environmental policy making.[21] The new department took over most of the environmental responsibilities. The Department of Agriculture, however, remained in control of the nitrate policy formulation though it often consulted the new Department of the Environment (interviews, Department of Agriculture, 1993, Department of the Environment and Natural Resources 1993, LRF, 1993). Gradually, the Board of Agriculture gained more influence on nitrate policy making in agriculture. It was given responsibility for suggesting policy measures. Previously, the EPA was responsible for both proposing and evaluating policy measures and objectives. The changes meant that the EPA now had a general policy making function in nitrate policy making (interviews, EPA, 1994 and Board of Agriculture, 1994).

In 1990, nitrate pollution again reached the political agenda. Only two years after the government had tightened the nitrate policy, the EPA presented a research report which concluded that further measures were needed to limit nitrate pollution. The measures for which the EPA were calling resulted only in minor changes in the existing nitrate policy. The EPA acknowledged that new policy measures had to be within the limits of what was technically, economically and scientifically possible (Naturvårdsverket, 1990, pp. 71-2).

The administrative changes in the late 1980s implied that the Agricultural Board became a central actor in the nitrate policy process. It set up three committees to suggest specific policy measures which could meet the aims of the EPA report. These consisted of agricultural scientists and representatives from the Federation of Farmers, the EPA and the Agricultural Board. A compromise on tightening the existing regulatory policy instruments of the nitrate policy was reached. The committees recommended that specific rules on autumn crop cover should depend on regional conditions, that the areas in which the strictest rules on storage capacity and application of manure were in force should be extended, and that the authorisation of manure spreaders and fertiliser distributors should be a legal requirement (Lantbruksstyrelsen, 1990, 1990a, 1990b). With regard to autumn crop cover, the relevant committee recommended that farmers in Southern Sweden should be legally obliged to have 60 per cent autumn crop cover and that farmers in Northern Sweden should have 40 per cent (Lantbruksstyrelsen 1990, pp. 29-30). In the negotiations, farmers tried to persuade the Agricultural Board and the EPA to adopt rules for autumn crop cover which would take into account the conditions on each farm. The EPA and the Agricultural Board could agree to make rules based on the type of soil in local areas but the Federation of Farmers could not accept this, arguing that it was unfair to some farmers. Since farmers could not accept local rules on autumn crop cover, the EPA and the Agricultural Board agreed to recommend regional rules to simplify monitoring (Eckerberg, 1994, p. 91; Lantbruksstyrelsen, 1990, p. 29).

The government bill did not fully take into account the compromise reached in the three committees. It did not suggest any extension of the areas in which the strict rules on storage and application of manure should be in force. This concession to farmers was counterbalanced by a tax on liquid ammonia which was intended to give it parity with other types of chemical fertilisers. The *Riksdag* accepted the government bill (Proposition 1990/91: 90, pp. 437-46; JoU 1990/91: 30, p. 237). While farmers avoided the introduction of policy instruments which were totally unacceptable to them, the government tightened the general objective of the nitrate policy. Sweden had ratified international agreements which aimed at a 50 per cent reduction in the nitrate pollution of the sea.[22] To fulfil these agreements, Swedish agriculture had to reduce nitrate run-offs by 50 per cent by 1995. Previously, the objective was supposed to be achieved by the year 2000 (Proposition 1990/91: 90, p. 76).

Changes in regional administration weakened farmers' control over the implementation process. In 1991, an administrative reform merged the county agricultural boards and other regional administrative boards with the county administrative boards. Both the Agricultural Board and the Federation of Farmers opposed the reform, arguing that the current organisation was effective and, consequently, there was no need for changes. But they could not mobilise sufficient support to prevent the merger from taking place (Proposition 1988/89: 154), thus the Agricultural Board lost a great deal of control over the regional administration of the policies it administers (interview, Board of Agriculture, 1994). Therefore, implementing officials may give agricultural concerns lower priority in the implementation of the nitrate policy.

Conclusion: a comparison of the policy positions of agricultural authorities and of the contents of national nitrate policies

This examination has focused on the contents of nitrate policies and on the policy preferences of the actors involved in the nitrate policy process, in particular those of agricultural state actors. Swedish and Danish agricultural authorities had significantly different policy positions. While the Danish Ministry of Agriculture developed policy preferences which were very sympathetic to farmers' interests, Swedish agricultural authorities were much more careful not to pursue agricultural interests alone. The Swedish Agricultural Board reached compromises with the Environmental Protection Agency, leaving farmers isolated politically. These compromises could only be reached because they coincided with the bureaucratic interests of the Board. By accepting fertiliser taxes, the Agricultural Board was given new responsibilities in agri-environmental policy implementation, and thus secured for itself more bureaucratic resources. While Danish farmers chose a confrontational strategy in dealing with environmental interests, Swedish farmers were much more moderate. The Swedes must have recognised that they did not have sufficient political support within the state apparatus to pursue a successful confrontational strategy.

The Danish and the Swedish nitrate policy apply several similar policy instruments; however, there are major differences. Since 1984, the Swedes have employed fertiliser taxes which, so far, have been impossible to introduce in Denmark. The major policy instrument in the Danish nitrate policy is obligatory fertiliser and crop rotation plans and accounts - a highly individualised regulatory policy instrument.

Both Danish and Swedish farmers preferred this instrument to environmental taxes, but the Swedes did not succeed in achieving this. In both countries, the nitrate policy has applied environmental subsidies to increase manure storage facilities. However, the Danish level of support was higher and was granted for a longer period than in Sweden. In Sweden, the state covered 20 per cent of the enlargement costs at each entitled farm, the maximum amount being 25,000 Swedish *kronor* (£2,275).[23] The scheme was in effect from 1989 to 1991. In Denmark, the state covered between 25 and 40 per cent of the costs, the maximum amount being approximately 100,000 Danish *kroner* (£9,877) in the 1980s and 150.000 *kroner* (£13,465) in the 1990s. The scheme ran from 1987 to the end of 1994. Compared to Danish farmers, Swedish farmers bear a much larger share of the perceived economic and political costs of the nitrate policy.

Likewise, the choice of policy objectives was much more disadvantageous for Swedish farmers than it was for Danish farmers. While farmers in Sweden had to implement a 50 per cent reduction of nitrate run-offs by 1995, Danish farmers have five additional years to achieve this objective. More importantly, Danish farmers avoided the nitrate policy setting aims for the reduction in fertiliser consumption. In contrast, the Swedish nitrate policy set specific reduction targets.

Since the Swedish nitrate policy consists of more high cost instruments and disadvantageous objectives than the Danish, we can classify the Swedish nitrate policy as a high cost policy and the Danish as a low cost policy. These findings show that Danish farmers were more successful than their Swedish colleagues in transferring the main principle of agricultural policy making, the principle of state responsibility, to environmental policy making. Swedish farmers had to accept that the polluter pays principle was applied more strictly.

Notes

1. Excluding Greenland and the Faroe Islands.

2. In 1993, it changed its name to the Family Farmers' Association (*Dansk Familielandbrug*).

3. From 1971 until 1973 the official name of the ministry was the Ministry of Pollution Abatement.

4. Ochre run-offs are caused by drainage of wetland for agricultural purposes.

5. Phosphorus pollution was put down to industry and households (Miljøstyrelsen, 1984).

6. Although the agricultural advisory service recommended manure storage capacity for six months, only 37 per cent of the farms with animal production were able to fulfil the requirements in May 1985, when the *Folketing* enacted

the NPO action plan (De Danske Landboforeninger og Danske Husmandsforeninger, 1989, pp. 4-5).

7. The Agricultural Commission based its investigations of the nitrate problem upon an earlier report prepared by the Ministry of Agriculture. The report claimed that the nitrate pollution in agriculture was not as serious as claimed by the EPA (Landbrugsministeriet, 1984).

8. The fact that the so-called confidential report had been published in an agricultural journal six months earlier shows that the decision making process was chaotic (Andersen and Hansen, 1991, p. 107).

9. Catch crops are crops which are planted simultaneously with the main crop. After harvest of the main crop, the catch crops continue to grow and during the winter they cover the field. They take up nitrate and thus limit nitrate run-offs.

10. A year later, a report from the EPA stated that the reduction had been less than 20 per cent (Miljøstyrelsen, 1992, p. 85).

11. By assigning the responsibility for nitrate policy making to the Ministry of Agriculture, the government followed the recommendations of the World Commission on Environment and Development, chaired by the Norwegian prime minister Brundtland. The Commission recommended that environmental concerns be integrated into sectoral policies (Den danske Regering, 1988, pp. 13, 20).

12. Generally, the Society is much more successful in agenda setting than in policy formulation (Rehling, 1994, p. 91).

13. The national subsidy scheme was abolished in 1993, but replaced by EC agri-environmental measures which were introduced in the 1992 agricultural policy reform.

14. The Swedish case refers to interviews undertaken in 1993. These interviews were conducted by Svante Forsberg, Department of Political Science, Gothenburg University, Sweden.

15. The sources do not reveal any direct relation between the entitlement to environmental subsidies and alteration of the rules on environmental permissions. Since the Federation of Farmers did not resist the change of the latter (*Land* 1972, no. 44), I believe that the two situations were interrelated.

16. This conclusion is supported by the fact that in 1972 and 1973, the farmers' section of LRF's newspaper *Land* had only a few articles on environmental

protection. In *Land's* consumer section, Micheletti (1990, p. 107) found '[n]umerous articles on the environment and chemical pesticides' in the early 1970s. Therefore, farmers seemed to be more concerned about mobilising new political support than about protecting the environment.

17. The county agricultural societies (*Hushållningssällskapen*) which are independent, private organisations also advise farmers. The relative importance of the state advisory service is evident when one compares the number of employees in the advisory services. While the county agricultural boards employed 1,130 advisors (1,020 man years) in 1990, the county agricultural societies employed only about 200 in 1991 (140 man years are used for advising in agriculture) (SOU 1990: 87, 17; SOU 1992: 99, pp. 20, 59-60).

18. Intensive farming is a main cause of nitrate run-offs.

19. Critique from the Federation of Farmers is almost absent. The enactment of the government bill is only addressed in one issue of its newspaper *Land* in which it was emphasised that the bill was a compromise between environmental and agricultural interests. The fertiliser tax was the only measure being criticised (*Land* no. 9 1988).

20. See note above.

21. From 1993 until 1987, the environmental division of the Department of Agriculture was more than twice as large as the agricultural division.

22. The agreements are the Second International Conference on the Protection of the North Sea and the Nordic Action Plan Against Pollution of the Sea (Nordisk Råd, 1993, pp. 7-8, bilag 1).

23. 1990 exchange rate.

6 Agricultural policy reforms in the European Community and Sweden[1]

Introduction

The meso-level model developed in chapter 2 points out that coalition building between farmers and state (or European Community) agricultural authorities is an important variable affecting the outcome of policy. To understand policy choices in the European Community's 1992 agricultural policy reform and in Sweden 1990 policy reform, we need to examine the content of the policy changes and to reveal the pattern of coalition building in the two cases. The differences in policy content are analysed by applying the classification of policy reforms based on a slightly modified version of Peter Hall's distinction between first, second and third order reforms (see chapter 4). In first order reforms, policy instrument settings are changed. Instruments, policy objectives and policy principles remain the same. Second order reforms are characterised by alterations in instruments and objectives. Policy principles are maintained. Third order reforms are much more fundamental: instruments, objectives and policy principles are all changed. To reveal the coalitions in the reform process, the theoretical policy network model suggests that we undertake an analysis of the actor's policy positions. In particular, the analysis should focus on the agricultural authorities' policy preferences. As the model points out, such actors' positions cannot be explained without paying regard to the structure of agricultural policy networks. The major question of this part of the empirical analysis is: how sympathetic were Swedish and EC agricultural authorities to the interests of farmers during the reform processes? Chapter 7 takes a closer look at the agricultural policy networks in order to establish whether differences in the policy positions of the state agricultural authorities can be explained by differences in network structures.

In most Western countries, agricultural policies went through their formative phases in the early 1930s. The Common Agricultural Policy (CAP) of the European Community was developed during the 1960s; however, the nature of the Common Agricultural Policy can be traced back to, particularly, the German agricultural

110

policy of the 1930s (Tracy, 1989, p. 269; Hendriks, 1994, pp. 59-61). Swedish agricultural policy was until 1991 very similar to the Common Agricultural Policy. Both policies were *high price policies* in which high consumer prices were the mechanism subsidising farmers. By restricting imports through variable levies and by internal market manipulation, consumer prices were kept at a higher level than if free market forces prevailed. Market manipulation works as follows: when supplies increase and prices fall below a politically decided level, public authorities intervene in the market. They buy up surplus products, stock them up and they subsidise exports. Consequently, consumer prices in the internal market stabilise. When supplies decrease and prices reach the maximum level, public authorities sell out stocks and imports increase as they become profitable. From the early 1970s until the mid-1980s, Swedish agricultural policy differed slightly from this model. Food subsidies were applied to keep down consumer prices on some basic foods (SOU 1977: 17, pp. 177-83; SOU 1984: 86, pp. 78-81). *Low price policies* are based upon another principle. They do not intervene into the market. Farmers receive the world market price for their products. Instead of price support, the state pays direct subsidies to farmers (see Kjeldsen-Kragh, 1986 for a comparison of the two systems).[2]

Until the reforms in the early 1990s, both the EC and the Swedish agricultural policy were based on the principle of state (or EC) responsibility for imbalances in agriculture which protects farmers from free market forces. As we shall see, farmers were happy with this principle. It protects a highly valued way of life by providing economic benefits, and it put them in a privileged position in the agricultural policy process. Hence, farmers will defend the existing agricultural policy. Agricultural policy reformers based their policy proposals upon market principles, holding that the state should cease to protect farmers from the market. Thus, agricultural reform processes involve a clash between two conflicting policy principles. Since the introduction of more market forces into the agricultural sector violates the principle of state (or EC) responsibility and, consequently, creates uncertainty and income instability, farmers will oppose agricultural policies based on the market principle.

To a very large extent, the Common Agricultural Policy and Swedish national agricultural policy experienced similar types of policy problems from the late 1970s until the early 1990s. Surplus production has been the major problem of both; in fact, it has developed into a chronic problem. Despite warnings from the OEEC (later OECD), from the Economic Commission for Europe (ECE) and even from the first EC agriculture commissioner, Sicco Mansholt, the EC adopted a high price policy in 1962. In five reports, published between 1955 and 1961, the OEEC stressed the dangers of a high price system and pointed out that it was a dubious instrument with which to increase farmers' income. The Economic Commission for Europe predicted as early as in 1960 that major surpluses in dairy products, wheat and sugar would arise (Tracy, 1989, pp. 237-8; Neville-Rolfe, 1984, p. 198). That was exactly what happened. In 1968, the European Community 'was well above self-sufficiency in several commodities, in particular milk, milk products, sugar and wheat' (Tracy, 1989, p. 267). Since the 1950s, Sweden has also had surplus production of, particularly, cereals and butter (SOU 1977: 17, p. 131). High price policies create incentives for farmers to intensify production; that is, they can use more production inputs than would be economically rational in situations in which a low price policy or free market forces prevailed. Therefore, high price policies create an imbalance between production and consumption; that is, artificially high prices stimulate farmers to produce more than consumers demand at these prices. This is a major reason for surplus production.

Surplus production burdens public budgets: firstly, stocking spending increases and, secondly, disposal of surplus products in the world market demands considerable export subsidies. Budgetary problems were not the only problem of the EC agricultural policy. Disposal of surpluses in the world market created international trade problems. The United States raised this issue, insisting that agriculture should be incorporated into the GATT Uruguay round which started in 1986 (Guyomard et al., 1993, p. 252).[3] Previous GATT rounds had dealt with agriculture but the agreements made were limited in character. In particular, the European Community's subsidised food exports were criticised by the Americans who were unwilling to conclude the Uruguay round unless there was also a deal on agriculture. High price policies motivating intensive farming also cause environmental damage and pollution, for instance pesticide and nitrate pollution. Finally, high price policies allocate the largest share of the subsidies to large farmers, while those most in need of support - the small farmers - receive only a limited share. For instance, in the EC, 20 percent of the farmers receive 80 per cent of the price support (Kommissionen, 1991, p. 2).

The institutions of the EC are unique and unlike any other national and international institutions. According to the Treaty of Rome (EC's 'constitution'), the Commission proposes, the Parliament advises, the Council of Ministers decides, and the Court interprets. However, in practice, the Council has played a significant role in policy initiation (Nugent, 1991, p. 101). There is one council for each of the Commission's policy areas. They consist of national ministers who are responsible for the particular policy area, for instance agriculture. The Council of Foreign Affairs Ministers deals with general issues and occasionally it handles cases which have deadlocked in the specialised councils. Heads of government (or state) meet in the European Council.

The special features of the Swedish administrative organisation are briefly described in the introduction to chapter 5.

The European Community

Although the EC had surplus production in the dairy, sugar and wheat sectors as early as 1968, it was not until 1977 that something was done about the underlying problems of the Common Agricultural Policy when the Community introduced a co-responsibility levy on milk. The idea of the levy was that dairy farmers should bear some costs of surplus production. When the production exceeded a certain threshold, prices were reduced the following year. However, the reductions were not automatic. The impact of the co-responsibility levy was very limited since the Council of Agriculture Ministers offset the effects by increasing prices enough to level out the levy. Guarantee thresholds were introduced in the arable sectors in 1981. These measures were intended to penalise farmers by lowering prices when production reached a certain level. However, like the co-responsibility levy, the guarantee thresholds were not automatically put into effect and, consequently, their impact was very limited (Moyer and Josling, 1990, pp. 60-3, 85; Kommissionen, 1985, pp. 19, 25; Tanner and Swinbank, 1987, p. 292).

In 1983, rapidly rising spending in the dairy regime caused a severe budgetary crisis. Policy makers had to choose between a collapse of the Common Agricultural Policy or a change in the dairy regime. The Agricultural Directorate considered at least three alternatives. One was a large cut in milk prices. This alternative did not win support because it would have had unacceptable effects on farmers' income and would only have reduced production in the longer term. Another alternative was to adopt a large increase in the co-responsibility levy but the Commission rejected this on similar grounds. Milk quotas had more support, although policy makers would rather be without them. Such a measure would reduce production immediately and farmers' income would not be too badly affected (Moyer and Josling, 1990, pp. 70-1; Kommissionen, 1985, p. 19; Tanner and Swinbank, 1987, pp. 292-3). Although farmers would clearly have preferred that the existing milk regime continued, they found quotas the most acceptable solution to the budgetary crises (Moyer and Josling, 1990, p. 70). According to Gardner (1987, p. 171), the European Farmers' Association, COPA, even persuaded the Commission and the Council to adopt milk quotas. The quota scheme was approved by the heads of government in March 1984.

By the mid-1980s, overproduction in the cereals sector had also become a major problem of the Common Agricultural Policy. The Commission's 'green paper', published in 1985, pointed this out. The fundamental problem was that production increased by 1.5 to 2 per cent annually while consumption increased by only 0.5 per cent annually. To find outlets for the surplus production, the EC subsidised cereals exports to the world market. The Commission realised that this practice was likely to create problems with the EC's trading partners. Further, the 'green paper' acknowledged that intensive farming had caused environmental damage and pollution. Interestingly, the Commission stated that the polluter pays principle should apply in agriculture; however, the Commission totally watered down the principle, saying that agri-environmental protection is a public good which farmers should be paid to deliver (Kommissionen, 1985, pp. 50-1). This is a clear example of the transfer of an agricultural policy principle to environmental policies.

Since the 'green paper' had already considered the problems of the CAP, it could not come as any surprise that a new budgetary crisis was imminent. It happened in 1988 and threatened to bankrupt the EC. The problems were caused by rapidly increasing spending in the arable sectors, especially in cereals. At the same time, farmers' income was declining. Thus, measures coping with the budgetary problems also had to take farm incomes into consideration. The solution was budget stabilisers and voluntary set aside schemes which were seen as the least controversial alternative to cope with the budgetary crises. The stabiliser package included a maximum guaranteed quantity (MGQ) which was to be fixed every year. If the MGQ was exceeded one year, a price cut, determined in advance, would automatically be put into effect the following year. In a year of overproduction, a co-responsibility levy would be collected in advance. It would only be refunded if MGQ was not exceeded (Moyer and Josling, 1990, pp. 78-98). However, the budget stabilisers were not sufficient to avoid yet another budgetary crisis three years later.

The dynamics of the 1992 reform process

What was the driving force behind the 1992 agricultural policy reform? Was it the GATT negotiations, yet another threatening budgetary crisis or was it environmental concerns?

Although Community spokespersons officially denied any link between the Uruguay round and the reform process, there is no doubt that the two processes were interrelated (Ingersent et al., 1994, p. 77; Larsen, 1992, p. 6). As Kjeldahl (1994, p.5) points out: 'The nature of the so-called Mac Sharry reform was dictated not only by internal constraints but also by the urgent need to revitalise the GATT negotiations.' A civil servant in the Danish Ministry of Agriculture who was in a very good position closely to follow the events during the reform process supported this view. He told that some senior officials of the Agriculture Directorate returned from the GATT negotiations realising that to achieve an agreement the EC had to cut agricultural prices and use direct payments to support farmers' income (interview Ministry of Agriculture, 1993c). Further, the Dutch minister of agriculture who was the Agriculture Council president in the second half of 1991 made the link between a CAP reform and the GATT talks clear when he told the press: 'What is necessary for the Uruguay round is that ... there is an indication of the direction in which the CAP will be adjusted and reformed' (*AE* no. 1448, 12 July 1991, p. E/6). However, the agriculture commissioner, Ray Mac Sharry, had an important incentive to keep the GATT talks and the reform process separate. He did not want to give the impression that the CAP reform was triggered by American pressure (*AE* no. 1424, 25 January 1991, p. E/3). The impression that American pressure was the root cause of reform would not help attempts to persuade farmers and member states, in particular France, to accept it.

The GATT negotiations were not the only problem which the CAP faced in 1991. By the end of 1990 the recurrent problem of the policy - a forthcoming budget crisis - returned once more to the agricultural agenda. The Commission estimated that the agricultural spending in 1991 would increase by approximately 8 billion ECUs, 'the highest annual increase ever recorded' (Manegold, 1991, p. 119). A further 20 per cent increase was expected in 1992 (*AE* no. 1427, 15 February 1991, p. P/1). German reunification only accounted for about one-sixth of the increase (*AE* no. 1424, 25 January 1991, p. P/1);[4] five-sixth of the increase resulted from surplus production. Expenditures would increase by 32 per cent in the cereals sector, 20 per cent in the oilseed sector, 20 per cent in the dairy sector, 50 per cent in the beef sector[5] and sheepmeat spending would go up by one-third (Manegold, 1991, p. 119).

114

Unless the Community severely limited increases in spending, the ceiling of the agricultural guideline which the heads of government imposed in 1988 to control agricultural spending would be breached (*AE* no. 1422, 11 January 1991, p. P/1; Tracy, 1989, p. 309), thus causing a budgetary crisis. However, a budget crisis is perhaps a necessary, but not a sufficient, condition for reform of the Common Agricultural Policy. Harvey (1994, p. 235) points out that '[h]istory suggests that a mild overshoot [of the guideline] will lead to changes in the budgetary guideline rather than the policy.' When sufficient resources have been provided 'to deal with the [budget] crises, pressure is removed to take any further action' (Moyer and Josling, 1990, pp. 210-11).

Environmental concern has been mentioned as an important driving force behind the reform. As early as in February 1990, the Commission's environmental spokesman said that 'environmental policy must more and more become a determining factor in all CAP decisions' and that the polluter pays principle should be applied where possible in agriculture (*AE* no. 1378, 2 March 1990, p. E/1). Later that year, the Environment Council of Ministers and the environment commissioner issued a statement in favour of revising the Common Agricultural Policy in order to improve the environment (*AE* no. 1408, 28 September 1991, p. E/6). Integrating environmental measures into the Common Agricultural Policy was the aim of most environmental interest groups lobbying in the reform process (*AE* no. 1457, 13 September 1991, pp. E/6-E/7; no. 1424, 25 January 1991, pp. E/4-E/5; no. 1468, 29 November 1991, p. E/4;). However, environmental concern was not the major driving force behind the reform. This was evident when the environment commissioner in February 1991 attacked the CAP, claiming that it was an 'ecological failure' which had caused as much environmental damage as the polluting industries in Eastern Europe (*AE* no. 1477, 7 February 1991, p. P/1). This strong attack on the policy was an indication that he had been unable to influence the reform process. Even though the agriculture commissioner and the Agricultural Directorate legitimised the reform with environmental concerns (Kommissionen, 1991, pp. 10-11), it was not enough to satisfy the environment commissioner and the Environmental Directorate. In spite of serious attempts, the Environmental Directorate could not convince the Agricultural Directorate that environmental concerns should be a central element of the reform (interview, Ministry of Agriculture, 1993c). As a result, environmental benefits remained just a side effect. Limiting costly surplus production was the main objective. The set aside scheme which became a central element of the reform also had environmental purposes but according the Director General of the Agricultural Directorate 'environmental considerations were of secondary importance as far as set-aside was concerned; the real object was to reduce cereals production' (*AE* no. 1488, 24 April 1992, p. E/6; see also Landbrugsministeriet, 1992f; Grant, 1995, p. 12).

In conclusion, the GATT talks and the expected budget crisis, should nothing be done, were the main driving forces behind the 1992 policy reform. Probably neither of these two factors would have been strong enough to cause major alterations in the policy by themselves. The combined effect of the two was what eventually led to reform.

The EC's GATT proposal and the GATT breakdown in 1990

There were probably only very few EC agricultural policy makers who in 1986 could, or indeed would, foresee that the recently started GATT Uruguay round would six years later lead to the most far-reaching CAP reform ever. Nevertheless, EC agricultural policy makers had to accept that the EC's trading partners would no longer accept the CAP's distortions in the world market. The United States and the Cairns Group[6] had a strong belief that the principle of comparative advantages should prevail in international farm trade. Nor were most of the EC member states willing to sacrifice the gains which the EC had won in the other 14 issue areas of the Uruguay round because of agriculture (Manegold, 1991, p. 118; Nedergaard, 1994, p. 87; 1995, p. 118; *AE* no. 1405, 7 September 1990, p. E/7; no. 1413, 2 November 1990, p. P/1). EC commissioner Sir Leon Brittan clearly expressed the EC's interests in concluding the GATT talks when he said: 'the European Community has perhaps the strongest interests of anyone in a conclusion of the Uruguay Round' (*AE* no. 1460, 4 October 1991, p. E/3).

In 1986, the Uruguay round began in Punta del Esta, Uruguay. The ministerial declaration of the meeting stated that with respect to agriculture the aim was to liberalise world farm trade and to make it more orderly and predictable. The agenda clearly reflected US priorities (Rayner et al., 1993, pp. 63, 79) but it was also a reflection of the interests of the Cairns Group, a group of 14 'fair trading' nations, which had been formed at a meeting in Cairns, Australia the year before (Tracy, 1989, p. 351). Australia and New Zealand were the principal countries of this group. The United States and the Cairns Group 'shared a common purpose in seeking reforms that would ensure that agricultural trade and export shares be shaped more in the future by underlying comparative advantage than has occurred in the past' (Rayner et al., 1993, p. 92). Since the establishment of the Common Agricultural Policy in the 1960s, these countries had, to a large extent, been excluded from the EC market; a situation which the United States had tolerated as long as the EC was not a significant cereals exporter in the world market. The United States had subsequently become very critical of the CAP since the EC had developed into a large cereals exporter and had built up a considerable oilseed industry which had won former American market shares in the EC (Guyomard et al. 1993, p. 257). From being largely excluded from the EC market, the United States and the Cairns Group members had also been forced to compete with the EC's directly subsidised exports in their traditional non-EC export markets (Mahler, 1991, p. 35).[7] They could not accept this development and thus succeeded in including agriculture in the GATT round.

Throughout the first four years of GATT negotiations, the top priority of the EC was to ensure the continuation of a Common Agricultural Policy which could still be based upon its existing principles. Agriculture commissioner Ray Mac Sharry made this clear in July 1990 when he said that '[w]e are fully engaged in the Uruguay round process. But let me make it clear, we are doing so on the basis of our commitment to the CAP' (*AE*, 1990, no. 1398, p. P/2). Mac Sharry was against liberalising world agricultural trade, arguing that the whole idea of agricultural trade liberalisation was based on unreliable academic arguments (ibid, p. P/3). EC Commission president Delors supported him by claiming that the abolition of EC export subsidies would 'destroy the EC agriculture industry' (*AE*, no. 1400, 3 August 1990, p. E/2). This view was mirrored by a senior Commission official who said: 'If we try to apply the same principles to agriculture as for other types of industry, the result will be a bloody failure' (*AE*, no. 1398, 20 July 1990, p. P/3). Therefore, the EC Commission and, in particular, the agriculture commissioner and the Directorate for Agriculture (DG VI) had no intention of accepting the United States' and the Cairns Group's liberalisation demands just before the Uruguay round entered its final stages. The Commission would not accept that a future CAP would be based on the principles of the market place. Instead, there was strong support for the idea that the EC should continue to take responsibility for the well-being of the agricultural sector. In other words, the principle of state responsibility should still apply. This principle safeguarded farmers' basic interests by maintaining an agricultural policy which protected them from, particularly, US competition.

To prevent the breakdown of the Uruguay round at the final GATT conference in Brussels in December 1990, the agriculture commissioner presented a fairly radical GATT offer. He suggested a 30 per cent cut in internal support for the main commodities and a 30 per cent reduction in border protection, transferring variable import levies into fixed tariffs. Further, there would be a match in export subsidies. In accordance with the Punta del Esta ministerial declaration, the reductions would be calculated on the basis of the 1986 support level (*AE* no 1407, 21 September 1991, pp. P/3-P/8). The GATT offer caused significant disagreement among the commissioners. A minority group led by the commissioner for External Affairs, Frans Andriessen, who had the overall responsibility for EC in the GATT negotiations, argued that to meet the latest US demands for a reduction in support by 70 per cent, the EC had to go further than 30 per cent. Mac Sharry did not succeed in gaining a majority for his proposal from the 17 commissioners (*AE* no. 1407, 21 September 1990, pp. P/1, E/1). Despite this, he refused to alter his proposal because Andriessen's and other commissioners' attempts to 'Americanise' the proposal 'would have been too damaging for European farmers' (ibid., p. E/1). Two weeks later, the Commission agreed on the GATT offer. To obtain the acceptance of the external affairs commissioner, Mac Sharry had to accept minor alterations (*AE* no. 1409, 5 October 1990, p. E/1). The result of the dispute between the two commissioners was that the agriculture commissioner and the Agricultural Directorate ensured control over the formulation of the Commission's agricultural policy position in the GATT process.[8] The external affairs commissioner, who was more interested in reaching an overall GATT agreement with all its benefits rather than protecting the narrow interests of European agriculture, had not been able to influence the Commission's policy position to any significant extent.

European farmers opposed policy reforms. In Britain, the National Farmers' Union (NFU) was 'shocked' that the Commission could suggest such a large reduction in subsidies, stressing that the 30 per cent reduction was 'the very worst'. COPA, the European association of national farmers' unions, did not like the proposal either. It called on the Commission to maintain the existing high price policy and to ensure that the final GATT agreement included mechanisms compensating farmers for fluctuations in currency exchange rates or in world market prices (*AE* no. 1407, 21 September 1990, p. E/2). The Commission's proposal would, according to COPA's president, 'lead to the destruction of the CAP and of European agriculture' (*AE* no. 1410, 12 October 1990, p. P/2). European farmers were, thus, opposed to more market forces in agriculture.

After the Commission had agreed on the GATT offer, the Council of Agriculture Ministers discussed it. Traditionally, the Agriculture Council has been more indulgent to farmers' interests than the Commission. From 1968 until 1990, the Council has, in far the most cases, given farmers higher price increases than the Commission has proposed; only once did the Council adopt lower prices and in five cases, there were no differences (Fearne, 1991a, p. 108). Similarly, the Council decision on the GATT proposal was more sympathetic to the interests of farmers than the Commission's proposal. However, the Council had difficulties in reaching a compromise. Britain, the Netherlands and Denmark accepted the Commission's position while the others opposed it, with France, Germany and Ireland as the strongest critics (*AE* no. 1410, 12 October 1990, p. E/4). The opponents would not accept the proposal unless the Commission guaranteed that it would 'reorientate the support given to farmers in a suitable way ..., thereby ensuring appropriate levels of income support' (*AE* no. 1413, 2 November 1990, p. E/4). Eventually, the Council accepted the offer of the 30 per cent reduction, but, in return, the Commission issued a declaration emphasising that it would take care of farmers' interests after the conclusion of the Uruguay round. The Commission assured that 'the total level of assistance to the less-favoured regions should not be reduced as a result of the implementation of the outcome of the Uruguay Round' (*AE* no. 1414, 9 November 1990, p. E/1). Furthermore, the Commission had to specify for the member states how farmers' incomes could be protected once a GATT agreement was reached (*AE* no. 1412, 26 October 1990, p. E/4). Commission president Delors even told farmer representatives that no farmers would go out of business because of the GATT proposal (*AE* no. 1416, 23 November 1990, p. E/3). The Community's proposal would involve a second order policy reform; that is, there would be an alteration in the policy instruments but the principles on which the policy is based would remain exactly the same.

COPA and the COGECA (the European cooperatives' committee) were still not satisfied with the GATT proposal. In a common statement, they claimed that the Community's GATT offer was 'a threat to agriculture and the entire economic and social fabric of rural society' and, thus, condemned the proposal. COPA and COGECA could not accept changes in the existing policy, saying that the EC's liberalisation ambitions were going too far. They did, however, welcome the package of accompanying measures which would be put into effect to help those farmers a GATT deal would hit hardest. These measures should be drawn up within 'the framework of a price and market policy which continues to form the main source of income of EC farmers' (*AE* no. 1415, 16 November 1990, p. E/4). European farmers did not accept the second order reform which the Community's proposal would involve. They wanted things to remain unchanged. Consumers, on the other hand, argued that the Community's GATT proposal did not go far enough in trade liberalisation. The International Organisation of Consumer Unions (IOUC) and the EC consumer association, BEUC, released a joint statement in which they said that the EC's GATT offer was 'progress of a kind; but nothing like enough' (*AE* no. 1417, 30 November 1990, p. E/3). Instead of a 30 per cent cut in agricultural support, they suggested 50 per cent. However, the two consumer unions accepted that the EC should pay direct subsidies to farmers in order to meet social and environmental objectives, but they could not accept that these payments were linked to production output (ibid.).

The GATT conference in Brussels in December 1990 was planned to conclude the Uruguay round; however, the conference ended in a temporary breakdown (Manegold, 1991, pp. 114-18). Although negotiators had made considerable progress in most of the 15 issue areas, including the difficult textile question, they could not, after four years of hard negotiations, reach a compromise in agricultural trade (Atkin, 1993, p. 141; Mahler, 1991, p. 38; Rayner et al., 1993, p. 88). There were still major differences between, in particular, the United States and the EC (Manegold, 1991, p. 114). The US, as well as the Cairns Group, did not find the EC's GATT offer far-reaching enough to base the talks upon (Guyomard *et al*, 1993, p. 254). Further, the two trading partners were 'unable to accept the EC's refusal to offer specific commitments to improving import access and reducing export assistance' (Rayner et al., 1993, p. 88). However, one could also blame the United States for unwillingness to compromise (Manegold, 1991, p. 114).

The early reform proposals

Despite being rejected by its trading partners, EC's GATT proposal paved the way for the reform process which eventually led to reform of the Common Agricultural Policy in May 1992. Before the Council finally decided to reform the policy, the ground had to be prepared. The agriculture commissioner came to realise that there was no fast road to reform as neither the other commissioners nor the ministers of agriculture were prepared to accept the fast track.

In December 1990, the Agricultural Directorate had produced a paper outlining a reformed Common Agricultural Policy which was later leaked to the press. The centrepiece of the paper was a reform of the arable sector, cutting the guaranteed minimum cereals and oilseed prices by 45 per cent. To compensate farmers for income losses, a direct compensation scheme was to be introduced. Farmers with up to 30 hectares of cereals, oilseed and protein crops (reform crops) would be fully compensated, whereas compensation would be reduced gradually for larger crop growers. To qualify for direct aid, farmers had to set aside land. For farmers with less than 30 hectares of reform crops, there would be no obligation to set aside. Farmers with more than 30 hectares of reform crops were to set aside 25 per cent of the area exceeding 30 hectares and 35 per cent of the area over 80 hectares. 'For ecological reasons,' the paper stressed, 'the temporary fallow would be organised as a rotation of surfaces in order to rest the soil.' The Commission wanted to maintain the milk quota regime, which originally was temporary, but large milk producers would have their quotas cut by 10 per cent. The minimum price for milk was to be reduced by 10 per cent. Small milk producers would receive a direct compensation for the first 15 cows. Beef prices would be lowered by 15 per cent. Beef producers would receive increased direct payments for the first 90 cattle to compensate for price reductions. Both the changes in the milk and beef regimes aimed at extensifying the sectors. Farmers could only receive direct compensation if they had no more than one livestock unit per hectare. The changes proposed in the sheepmeat market regime also aimed at extensification. There would be an upper limit for the number of ewes for which the farmer could obtain a premium and there would be certain extensification criteria to meet to qualify for support. The Agricultural Directorate also suggested three accompanying measures to the reform. The EC would introduce a system of aids to encourage farmers to use environmentally friendly production methods, including a long term set aside programme. Farmers would be compensated for their income losses if they participated in the environmental scheme. Also included was an early retirement scheme and income aid for farmers in less favoured areas. Reforming the Common Agricultural Policy as outlined by the Agricultural Directorate changed the relation between input prices and product prices, resulting in extensification of farming and would, thus, lower production. This benefits the environment. Basing a larger share of farmers' income upon hectare aid increases farmers' income certainty and stability and secures against drastic income losses, e.g. bad harvests (AE no. 1423, 18 January 1991, pp. E/2-E/7). The proposal represented a second order policy reform in the arable sector while the animal sector was subject only to a first order reform. What distinguished the proposal from a third order reform was that the principle of state responsibility remained the basic policy principle of the Common Agricultural Policy. Increasing income certainty and stability even strengthened the principle. Farmers would still receive large subsidies but in a different way. There was no serious attempt to leave the farmer to the mercy of the market. Therefore, the Community would continue to safeguard farmers' basic interests.

The reform paper triggered a debate within the Commission. Mac Sharry had 'the support of a solid core of commissioners - Bangemann [Germany], Andriessen [the Netherlands], Brittan [the UK], Christophersen [Denmark] and Millan [the UK]' (AE no. 1423, 18 January 1991, p. P/1). Commission president Delors led the opposition. He did not like the precise form of the proposal, but wanted a generalised, imprecise document (ibid., pp. P/1-P/2). Eventually, he persuaded Mac Sharry to prepare a loose reform paper. All the commissioners agreed that the policy had to be changed but there was, however, considerable disagreement on graduated compensatory payments with the British, Dutch and Danish commissioners being opposed to it (AE no. 1424, 25 January 1991, p. E/3). The agreement was thus based more on the need to reform the policy rather than on how to reform it.

European farmers disliked the way the Commission conducted this early stage of the reform process. COPA complained that the Commission had not consulted it before issuing reform statements. It repeated its traditional position which stated that 'price and market policy must remain the main source of farmers' income' (AE no. 1420, 21 December 1990, p. E/6). In Britain the National Farmers' Union complained that 'the ideas in the leaked Commission paper would devastate British farming. The measures in the paper are completely unacceptable' (AE no. 1423, 18 January 1991, p. E/8). One reason farmers opposed Mac Sharry's reform paper because a shift from a high price to a low price policy with direct payments would make them more vulnerable politically. The costs of direct income support are very visible and, thus, open to political opposition (e.g. Rieger, 1996, p. 117). Although farmers were against Mac Sharry's proposal, it protected their basic interests but by new means.

The Commission's reform ideas were presented to the Agriculture Council as an orientation paper. The paper underlined that the fundamental problem of European agriculture was that between 1973 and 1988, annual production output increased by 2 per cent whereas consumption only increased by 0.5 per cent. As a result, there had been surplus production which, in turn, increased stocking costs and created international trade problems for the EC. Production related support (which characterises high price policies) had encouraged intensive farming methods, causing environmental problems. Another major problem of the Common Agricultural Policy was that 80 per cent of the total price support went to 20 per cent of the farmers. A basic objective of the policy is to improve farmers' standard of living. This objective had not been achieved, the Commission concluded (Kommissionen, 1991, pp. 2-3). The Commission stated:

> Unfortunately, there is nothing new in the above analysis. Essentially, it has already been carried out several times, such as in 1985 when the Commission, on the basis of the 'green paper' instigated comprehensive deliberations on the future perspectives of EC agriculture (ibid., p. 4, my translation).

Introducing budget stabilisers in 1988 did not attack the underlying problems of the policy (ibid., p. 8).

To overcome the fundamental problems of the policy, the Commission paper maintained the ideas of the leaked December paper; that is, reducing guaranteed prices and compensating for income losses by direct subsidies, provided that the farmer set aside a certain percentage of his land. Furthermore, the Commission suggested that the CAP introduced accompanying measures which would encourage farmers to engage in environmentally friendly farming (ibid., pp. 13-15). Clearly, the reform proposal aimed at ensuring that a larger share of the agricultural budget went to small and medium-sized farms (ibid., p. 12). Compared to the December paper, this version outlined the reform ideas in broad and imprecise terms. Nevertheless, the reform proposal involved a second order reform in the CAP's arable market regimes.

Although the Commission formulated its document in general terms, it did not receive a warm welcome anywhere. In the Agriculture Council of Ministers, the majority was against it; in particular Britain, Denmark and the Netherlands were strongly opposed. As the agricultural newsletter *Agra Europe* argued: 'All three have highly capitalised, relatively large and efficient farms and have little to gain from policies supposed to help the smallest and the least efficient' (no. 1426, 8 February 1991, p. E/1). Germany 'complained that the plan ... lacked real substance' (ibid.). The French were less clear in their opposition. On the one hand, they 'wanted more assistance for small farmers' and, on the other hand, they were 'strongly against any plan which might hit arable crop exports' (ibid.). Ireland and the Southern member states chose a wait-and-see position (ibid.).

COPA strongly opposed what it saw as a 'threat to the existence of very many family farms, the viability and competitiveness of cooperatives and hence that of rural regions in the Community' (*AE* no. 1427, 14 February 1991, p. E/4). The association repeated its preference for a market and price policy rather than direct aids (ibid., p. E/5). The Belgian Farmers' Union (the *Boerenbond*) was particularly worried about the future of specialised farmers who 'would see their chances of survival disappear suddenly' (ibid.). Italy's leading agricultural association, *Confagricoltura*, which represents larger farmers, did not like 'the underlying motivation nor the operational proposals.' It objected 'to the principle of an inverse relationship between the amount of the aid and the size of farm' (*AE* no. 1428, 22 February 1991, p. E/2). In Denmark, the Farmers' Union, which represents medium-sized and larger farmers, rejected the reform ideas and declared that the existing agricultural policy should be maintained (Landboforeningerne, 1991, p. 20). By contrast, the Smallholders' Union supported the main principles (Husmandsforeningerne, 1991a, p. 18; 1992, p. 19). Despite the cold welcome of the Commission's broad and imprecise version of its leaked December paper, Mac Sharry pointed out that the December paper was still alive (*AE* no. 1428, 22 February 1991, p. P/4).

In late May 1991, COPA presented its reform proposal. Not surprisingly, COPA liked the idea of subsidising environmentally friendly methods in farming (COPA, 1991, p. 2). In the Commission proposal, the scheme was voluntary and farmers were to be fully compensated for their income losses, so COPA had no difficulties with that proposal. COPA wanted to maintain the existing high price policy with its system of high internal prices, import levies and export refunds. To cope with the surplus production, COPA suggested the implementation of a voluntary supply management scheme and production controls in which an important element was the development of an 'effective and efficient' non-food policy. Farmers suffering from income losses caused by such policies should receive compensation. In addition, surplus production should be diminished by seeking new outlets and regaining markets taken over by substitutes. Imports of products competing with EC surplus products should be restricted and 'the Community should pursue an active and dynamic export policy for farm and food products' (ibid., pp. 5-8). Basically, COPA wanted to maintain the existing high price policy; however, it recommended the introduction of supplementary policy instruments. In contrast to the Commission, COPA was unwilling to accept a second order reform of the Common Agricultural Policy.

COPA's proposals did not gain much support within the Commission. Although Mac Sharry was pleased that COPA acknowledged the need to reduce production, he pointed out that COPA's suggestion on import limitations was unrealistic (*Raadsnyt* no. 22, 21 June 1991, p. 3). However, there was no sign that the Agricultural Directorate believed in supply management. Its deputy director general was not convinced that voluntary production restrictions would cope with surplus production. He later pointed out that the history of such measures had not convinced the Commission that they were workable. In addition, he argued that maintaining the existing high price policy was not the way forward because it depended on 'export refunds which are ever more difficult to defend internationally' (*AE* no. 1462, 18 October 1991, p. E/3). Thus, the COPA proposal offered too little to influence the Commission's forthcoming formal reform proposal. Furthermore, COPA probably acted too late.

BEUC, the bureau of European consumers' associations, also published reform recommendations in May 1991. Being highly critical of the existing high price policy, BEUC argued that '[t]oo often, money spent on the CAP creates problems which lead to demands for more money to be spent to solve the same problems.' It maintained that a radical policy reform would be the best way to serve consumer interests. By radical reform, consumers meant 'the exposure of EC agriculture to market forces.' The union favoured a decoupling of support from production. In that respect, the Commission did not go far enough since the idea of linking direct aid to historic yields and farm sizes would not totally decouple production and support (*AE* no. 1442, 31 May 1991, p. E/9). Consumers were not against agricultural subsidies. Earlier on, they had said that 'the social objectives implicit in the CAP should be rendered more explicit by operating through the Community's three structural funds - the social fund, the regional fund and the agricultural guidance fund' (National Consumer Council, 1989, p. 191). Since the guidance fund was the only one under control of the EC's Agricultural Directorate, such a recommendation was highly controversial within the Directorate and among farmers. The consumer position was, therefore, too radical to influence the content of the reform.

The EC's food and drink industry confederation, CIAA, also produced a reform paper which outlined the principles on which a Common Agricultural Policy reform should be based. Because the food processing industry is divided, its reform principles were formulated in broad and vague terms. The confederation could agree that agricultural prices should better reflect the balance between supply and demand, but there were no recommendations as to how this principle should be implemented. What is more, the confederation warned against the introduction of social policy instruments into the policy. This could have unacceptable effects on the supply of raw material to the industry and, thus, its competitiveness (*AE* no. 1445, 21 June 1991, pp. E/2-E/3). It is unlikely that the CIAA paper had much influence upon the reform process due to its vague and imprecise nature.

The Commission presented its formal reform proposal in July 1991. Basically, it was based upon the same ideas as the GATT proposal from autumn 1990, the leaked reform paper from December 1990 and the orientation paper presented to the Council in February 1991. The centrepiece of the proposal was a reform of the arable sector. The Commission still favoured a second order reform in the arable market regimes and a first order reform in the animal sector. On the one hand, the reform paper rejected COPA's proposal on supply management and import restrictions (Kommissionen 1991a, pp. 4-5). On the other hand, the Commission did not go as far as consumers may have wished.

Cereals prices were to be reduced to the expected world market level which meant a 35 per cent cut in EC prices.[9] Oilseed and protein crop prices would also be reduced to the world market level. Small producers of cereals, oilseed and protein crops (reform crops) would still be exempted from setting aside land and would be fully compensated for income losses. Farmers producing more than 92 tons of cereals (on average, this is produced on 20 hectares) were required to set aside 15 per cent of the land upon which they had previously grown reform crops. Compensation paid to these farmers would be proportionally reduced according to the size of the area in which the farmer grew reform crops. Milk quotas were to be cut less than suggested in the December paper as they were to be reduced by 5 per cent, but the price cut would still be 10 per cent. The milk price cut was not as radical as it may look at first sight. Prices were reduced mainly because of falling feeding costs which, in turn, was caused by the decrease in the prices on cereals, oilseed and protein crops. The Commission suggested direct aids to compensate for the quota cuts. Beef prices would fall by 15 per cent of which the 10 per cent was due to lower feeding costs. To compensate beef producers for the 5 per cent income loss, the Commission recommended an increase in direct payments for the first 90 cattle. The reform proposal maintained the idea of extensification; however, the criteria for receiving support for this purpose were considerably lessened. While in the December proposal extensification meant one livestock unit or less per hectare, exstensification was now 1.4 livestock unit in less favoured areas and two in other areas. In the sheepmeat sector, the Commission upheld the proposal for putting an upper limit on the number of headage premia and for furthering extensification. All these changes were to be implemented over three years. Accompanying measures consisting of an agri-environmental programme, an afforestation programme and an early retirement scheme were still part of the reform proposal. Farmers would receive compensations for income losses resulting from the agri-environmental and the afforestation programmes (Kommissionen, 1991a). The Commission argued that the reform would extensify production which, in turn, would reduce production output and benefit the environment (ibid., p. 15). The Commission's need to curb spending increases motivated the reform. However, the reform would, on average, increase expenditure by approximately 2.3 billion ECUs annually. Furthermore, the accompanying measures would cost about four billion ECU in the first five-year period (ibid., pp. 41-4). The Commission argued that the increase in spending was fully justified (ibid., pp. 3, 38) because by 1997 the expenditure would be even higher if nothing were done (ibid., p. 6).

From the preparation of the GATT proposal in late 1990 until the Commission's final reform proposal was presented in July 1991, the Agricultural Directorate was in control of the reform process. It had set up a small think-tank which designed the Commission's proposals (interview, Ministry of Agriculture, 1993c). The core of the think-tank seems to have involved only five people (*AE* no. 1446, 28 June 1992, p. P/4). However, the External Affairs Directorate, which had the overall responsibility for the GATT talks, did have some influence. It analysed reform papers prepared by the Agricultural Directorate to ensure that they did not violate the External Affairs Directorate's interest in concluding the GATT negotiations. The fact that the external affairs commissioner, Frans Andriessen, was the former agriculture commissioner and had a relatively strong position within the Commission probably helped to ensure some influence of the External Affairs Directorate (interview, Ministry of Agriculture, 1993c).

Most national farmers' unions opposed the reform proposal. French farmers organised in the FNSEA rejected it, criticising the Commission for not taking into account the views of farmer associations. The *Bauernverband* in Germany stated that price reductions would ruin many farmers. In Britain, the National Farmers' Union was concerned about income losses for, in particular, cereals growers and sheep farmers. In Denmark, the president of the Agricultural Council and the Farmers' Union said he was 'alarmed that the plan will require the most massive bureaucratic controls to function satisfactory.' He also criticised the reform plan for favouring small, unproductive farms in Southern Europe (*AE* no. 1448, 12 July 1991, pp. P/3-P/4, E/5). The Irish Farmers' Association also expressed strong opposition to the reform proposal (*AE* no. 1450, 26 July 1991, p. N/4). In Italy, *Confagricoltura*, which represents the large farms, also opposed the Commission's recommendations (*AE* no. 1460, 4 October 1991, p. E/2). However, not all farmers' unions opposed Mac Sharry's reform ideas. The Italian farm union *Confederazione Italiana Coltivatori*, which represents smaller farmers, even demonstrated in Brussels in support of the reform (*AE* no. 1448, 12 July 1991, p. P/3), and in Denmark, the Smallholders' Union remained in favour of the reform (Husmandsforeningerne, 1991a, p. 18; 1992, p. 19).

Basically, the negotiations in the Agriculture Council of Ministers did not alter the nature of the Commission's reform proposal. However, as on previous occasions, the Council was more sympathetic to agricultural interests than the Commission. The ministers did not lower the cereals intervention price from 155 ECUs/t down to 90 ECUs/t within three years which the Commission recommended, but only to 100 ECUs/t. Thus, the intervention system and the system of export subsidies and import levies remained intact. Although these elements of the old CAP were maintained, the reform still involved a basic change in policy instruments employed in the arable sector. In the animal sector, the Council adopted a first order reform.

Other elements of the Commission's proposal were also unacceptable to the

Council. The non-acceptance of the graduation of the set aside compensation in favour of smaller and medium-sized farms was an important alteration. Furthermore, the ministers succeeded in increasing the protection against non-EC cereals imports by raising the difference between the target and the threshold prices (*Agra Europe* special report no. 65, 1992, p. 5). Originally, the Commission proposed a difference of only 10 per cent (Kommissionen, 1991a, p. 9) but the ministers agreed to increase it to 41 per cent (*Official Journal*, L 181, Council Regulation (EEC) no. 1766 of 30 June 1992). France, Ireland and the Southern European member states had emphasised the need to increase the border protection. In the animal sector, the Commission's proposals were also modified. The Council refused to make any changes in the dairy regime except for a small cut in butter prices. Basically, the ministers accepted the Commission's recommendations on changes in the beef sector; however, they declined to adopt the ceiling of 90 animals which would limit payments of direct compensation to large farmers. Likewise, the Council accepted the Commission's sheepmeat proposal, but raised the upper limit for headage premia (*Agra Europe* special report no. 65, 1992, pp. 36-59). The Council adopted the Commission's measures to extensify production in the beef and sheepmeat sectors whereas they declined to introduce them in the dairy sector.

The majority of member states could, more or less, support the principles of the Commission's reform plan.[10] While most of them had objections to the price cuts, they supported the idea of giving full and permanent compensatory payments to farmers. The French position, however, was less clear. On the one hand, the French minister had to take care of the large and efficient arable exporters in Northern France and, on the other hand, he wanted to protect the incomes of the small farmers in the South. Germany's support for the reform was rather important for the breakthrough in the reform negotiations. The desire for a GATT deal changed Germany's traditional conservative position in the CAP process (*Agra Europe* special report no. 65, 1992, p. 3).

Belgium, Denmark, the Netherlands and the United Kingdom opposed the reform plans. They neither liked the Commission's radical price cuts nor were they pleased with the idea of giving full and permanent direct support to farmers. Instead, they argued that the settings of the existing policy instruments should be adjusted in the arable sector by cutting prices but without introducing new instruments for direct payments. In other words, they preferred a first order reform of the existing policy. Although there seemed to be agreement on the principles in the beginning of the Council process, it later turned out that there was considerable disagreement. The Dutch president of the Agriculture Council declared already in July 1991 that 'all delegations [i.e. members states, CD] agree that if there is no fundamental reform we will have an unbearable situation' (*AE* no. 1449, 19 July 1991, p. E/1). Two months later, he said there was general agreement in the Council that a move from production support to direct income support was necessary (*AE* no. 1459, 27 September 1991, p. E/1). However, he seems to have misjudged the situation. As late as March 1992, Belgium, Denmark, the Netherlands and the United Kingdom were still demanding a discussion on principles. When they realised that there was no chance of this, they joined the majority. As a result, the ministers were able to reach agreement on a second order reform in the Common Agricultural Policy on May 21, 1992.

In return for accepting the principles of the reform, Belgium, Denmark, the Netherlands and the United Kingdom obtained some concessions. Firstly, discrimination against large farmers was taken out of the reform. France had also argued for this and, thus, obtained an important concession. Denmark and the United Kingdom were given environmental concessions. In particular, Britain had argued for more environmental concerns in the set aside scheme. The two member states persuaded the others that set aside could be both rotational and non-rotational. If a farmer chose non-rotational set aside, the set aside percentage had to be higher than that for rotational. The percentage was to be decided later (*Agra Europe* special report no. 65, 1992, p. 8). The Commission opposed non-rotational set aside, arguing that allowing it would undermine the aim of the production reduction because farmers would set aside the least productive fields. The budgetary effects of the Council's alterations of the Commission's proposal were relatively severe. While the Commission had estimated that expenditures would increase by 2.3 billion ECUs annually, the Council decision would cost another three billion ECUs annually in the arable sector alone. This would demand an increase in the agricultural guideline by 1996 (*AE* no. 1497, 26 June 1992, p. P/1). This problem was left unresolved by the ministers of agriculture.

The farmer associations were not particularly pleased with the reform; however, some did accept it. COPA stated that the reform did 'not constitute a basis for a coherent Common Agricultural Policy capable of solving the most urgent problems currently facing the agricultural sector' (COPA, 1992). To emphasise that COPA had no part in the undesirable substance of the reform, the association made it clear that politicians 'should take full responsibility for the consequences that these decisions will have on farmers' (ibid.). Germany's Farm Union pointed out that the reform was 'disastrous' and in France the FNSEA called it an economic and political error. In Belgium and Italy, farmers were also opposing the agreement. Danish farmers were satisfied and Dutch farmers were moderately satisfied. The National Farmers' Union in Britain was pleased that large farms were also entitled to full compensation for price cuts. Irish, Spanish and Greek farmers were largely content with the reform (*AE* no. 1493, 29 May 1992, pp. E/1-E/3). BEUC (the European Bureau of Consumers' Associations) was not fully satisfied, saying the '[price] cuts are a welcome first step but they are far from the complete and integrated reform for which BEUC has been pressing for many years' (BEUC, 1992; see also National Consumer Council, 1995, p. 2).

Throughout the reform process, policy makers were mainly concerned with the policy instruments of the Common Agricultural Policy. The basic objectives of the policy remained unchanged but not wholly untouched. The Treaty of Rome states that the policy objectives of the Common Agricultural Policy are: to increase agricultural productivity, to ensure a fair standard of living for the agricultural community, to stabilise markets, to assure the availability of supplies and to ensure that supplies reach consumers at reasonable prices (De europæiske Fællesskaber, 1987,

127

pp. 253-4). Policy makers, especially within the Commission, stressed the need to fulfil the income objective in a more appropriate way (Rieger, 1996, p. 116). They favoured the introduction of policy instruments which could direct a larger share of the subsidies to the 80 per cent of farmers who had previously received only 20 per cent of the support (Kommissionen, 1991, p. 12). Environmental protection became a more valued aspect of the policy. Besides being food producers, farmers should also be custodians of the countryside (ibid., pp. 10-11). This aspect of the CAP was further strengthened during the Council negotiations in which, particularly, the British minister of agriculture argued for more green measures to be included in the CAP.

In conclusion, the reform of the Common Agricultural Policy consisted of a second order policy reform in the arable sector and a first order reform in the animal sector. European farmers were able to avoid a more radical reform because the agriculture commissioner and the Commission's Agricultural Directorate supported the basic principle of the CAP: the principle of EC responsibility for imbalances in agriculture. This principle implies that the EC should protect farmers from the market forces. The controversies between the Commission and farmers' associations were not about whether farmers should be subsidised or not, but centred around the way in which the EC should subsidise.

Sweden

Ever since the 1950s, Sweden has experienced surplus production of, particularly, cereals and butter (SOU 1977: 17, p. 131). Sweden has aimed at self-sufficiency in food production so that she would not be affected in war or crisis. In order to produce sufficient food stuffs in periods of scarce production input, the Swedes maintained agricultural production capabilities above the level required for peacetime production. Excess capacity in means of production resulted in increasing surpluses. The 1977 agricultural policy deliberately directed excess capability towards cereals production (SOU 1984: 86, p. 310). Therefore, later changes in the agricultural policy aimed at the cereals sector.

The early proposals to reduce overproduction of cereals were prepared by traditional agricultural policy makers. A state commission on food policy was set up in 1983 to evaluate the 1977 policy. It was commissioned to prepare proposals which could establish a balance between supply and demand. The measures to achieve this should hold farmers responsible for the costs of surplus production (SOU 1984: 86, pp. 29-38). The commission concluded that the state should subsidise the production needed for peacetime consumption. In principle, farmers should take financial responsibility for surplus production which could not profitably be exported; however, in practice, the state should bear parts of the costs. The commission recommended that the state paid 40 per cent of the costs while farmers' share was 60 per cent. During the five-year period to follow, the state's share of the costs was estimated at 600 million Swedish *kronor*. The commission's intention was that farmers should see to that the state's share could be abolished later. By increasing farmers' costs of overproduction, the commission hoped that they would begin to produce alternative crops, for instance energy crops (SOU 1984: 86, pp. 310-14). The farmers opposed these recommendations, criticising the commission's conclusions in a long minority statement in which they argued that the state alone was financially responsible for the surpluses (SOU 1984: 86, pp. 526). However, this argument did not attract sufficient support. The social democratic government and the *Riksdag* accepted these recommendations (Proposition 1984/85: 166, p. 47; JoU 1984/85: 33, pp. 34-5). After the *Riksdag* had approved the commission's proposals, the state agricultural boards and farmer representatives began to look for practical solutions to decrease cereals production.

In this light, a policy group on cereals (*spannmålsgruppen*) was set up to deal with surplus production. In June 1986, it submitted a report in which it presented proposals to decrease overproduction already in 1987. It suggested a voluntarily set aside scheme with economic compensation (Jordbruksdepartementet, 1986, pp. 44-6). The government approved the proposal (Proposition 1986/87: 146, pp. 9-10). In 1987, the policy group submitted another report suggesting increased use of the voluntarily set aside scheme and of economic incentives for the production of alternative crops (Jordbruksdepartementet, 1987a, pp. 37-40). These proposals were also adopted by the government (Jordbruksdepartementet, 1989, p. 94).

A committee on intensive farming (the Intensity Group) worked in parallel with the policy group on cereals. It investigated how farmers could diminish cereals surplus production and limit agri-environmental pollution[11] The Intensity Group concluded that the best way to lower the intensity of agricultural production and, thus, to reduce overproduction and environmental damage was to cut cereals prices considerably. Furthermore, it recommended that farmers should use more arable land for production of alternative crops. Discussing fertiliser taxes the group concluded that unless fertiliser taxes they were raised considerably, they would only have limited effect on production intensity (Jordbruksdepartementet, 1987, pp. 68-75). The report did not have any immediate consequences for the agricultural policy.

The dairy sector also produced more than the Swedes consumed. The government, therefore, introduced milk quotas in 1984 but abolished them in 1989. The GATT talks which had considerable influence on the EC agricultural policy process also affected Swedish agricultural policy, especially the dairy market regime. However, the impact was minor compared to the one on the EC. At the mid-term Montreal GATT meeting in 1988, the participants in the Uruguay round agreed to freeze the agricultural support linked to production. Hence, the Swedish government had to give direct headage payments to dairy farmers to fulfil the 1988 national price agreement. Farmers had traditionally opposed such policy instruments but forced by the situation they eventually accepted them (OECD, 1995, pp. 50-1).

In conclusion, there was not much change in Swedish agricultural policy in the 1980s. To cope with surplus production, the government introduced voluntary policy measures combined with economic subsidies. Consumers showed limited interest in the surplus problems as long as farmers had to take care of the problems. But passing on the costs to consumers captured their attention (SOU 1984: 86, p. 521). Not sur-prisingly, the policy measures coping with surplus production turned out to have only a limited impact. In the late 1980s, cereals surplus production was about 1.7-1.8 million tons. The total production was six million tons. Consequently, almost one-third had to be exported with state subsidies (Jordbruksdepartementet, 1989, p. 72).

The dynamics of the 1900 policy reform

As the analysis of the Common Agricultural Policy shows, the budgetary problem was a major driving force behind the Mac Sharry reform. Since Swedish agricultural policy also had surplus problems, one could expect that budgetary concerns drove the Swedish reform. However, this was not so. In 1988, agricultural subsidies con-sumed about 1 per cent of the state budget (Micheletti, 1990, p. 104). As the former minister of finance, Kjell-Olof Feldt, pointed out, the Department of Finance, which initiated the reform, was not particularly concerned with budgetary costs but with the indirect impacts of the agricultural policy; that is, welfare losses and inflation (Feldt, 1991, p. 343; see also Rabinowicz, 1993, p. 290).[12] A report published by the De-partment of Finance concluded that the agricultural high price policy was the reason the farm sector attracted economic resources which could be applied more efficiently in other sectors. Further, increasing food prices caused inflation (Finans-departementet, 1988, pp. 35-8). These macro-economic problems had the attention of the minister of finance who said in the *Riksdag*: 'Against the background of low fulfilment of the objectives [of the agricultural policy], an agricultural policy reform is needed so that it fulfils the objectives better, curbs inflation and reinforces economic growth' (Proposition 1988/89: 47, p. 10, my translation). A parliamentary working group motivated by similar concerns was set up to scrutinise the agricultural policy. It published a report which focused on a number of agricultural policy prob-lems.

A major problem with which the working group dealt was the macro-economic impact of food prices. Because rising food prices caused inflation, the issue became a major driving force of the reform as the above statement of the minister of finance indicates. Food prices increased more in Sweden than in other European OECD countries in the 1980s. While the average increase was 7.2 per cent in Europe, it was 9.7 per cent in Sweden. Swedish food price increases were well above the average increase in Swedish consumer prices which was 7.6 per cent. Part of the increase in food prices was caused by the abolition of food subsidies in 1984 (Jordbruksdepar-tementet, 1989, pp. 113-14). Food subsidies were introduced in the early 1970s to protect consumers from the costs of increasing commodity prices to farmers (Steen, 1988, p. 120). The parliamentary working stated that had food prices increased at the same rate as average consumer prices, the Swedish annual inflation rate would, it was estimated, have been 0.2-0.3 per cent lower. The lost growth in GDP was as-sessed to be 0.1-0.15 per cent annually which is equivalent to the loss of 2000-3000 jobs per year (Jordbruksdepartementet, 1989, p. 91). Perhaps, the group overestimated the agricultural policy's effects on the inflation rate. As Vail, Hasund and Drake (1994, p. 182) argue:

> We contend that the agricultural components of Swedish food policy actually received more blame for exceptionally high food prices than the facts justified. Farm receipts compromise only about one-fourth of retail food expenditure, and most of the 1980s food inflation can be traced to higher costs and profit margins in the food distribution system, where weak competitive forces and wage inflation had a major impact on retail prices.

Lack of competition in the food processing industry also attracted the concern of the parliamentary working group because it drove up prices, thus causing inflation and welfare losses (Jordbruksdepartementet, 1989, p. 115; interview, Board of Agriculture, 1994a). These consequences mobilised consumer interests (interview, Board of Agriculture, 1994a). Barriers to food imports allowed farmers' cooperatives to obtain a strong position in the Swedish food market. For instance, in the dairy market they controlled 100 per cent of the milk deliveries (Micheletti, 1990, p. 155). One result of the strong market position was that the food processing industry had little incentive to innovate. This caused problems with food quality and variety of supply (Jordbruksdepartementet, 1989, pp. 111-12).

The parliamentary working group also concluded that the high price policy had negative and unintended consequences within the farming sector. Firstly, most agricultural support ended up with farmers who had the best conditions to run efficient farms. They were, primarily, farmers in Southern and Mid- who have better climatic conditions, better soil and better farm structures than those in the North. Northern farmers were the ones who needed the support the most (ibid., pp. 159-63). Secondly, price support was capitalised. The idea of price support was to increase farmers' income; however, it never really worked as intended because price increases drove up the prices of farm land which, in turn, decreased the incomes of young farmers entering the business (ibid., pp. 127-33).

During the reform process, environmental concerns seemed to become a more important motive for reform. The parliamentary working group dealt with agricultural pollution and environmental damage. It acknowledged that specialised and intensive agricultural production caused environmental problems (ibid., pp. 134-51). An important aim of agricultural policy in post-war Sweden was to rationalise production. Specialisation, intensification and structural development were the means to make production more efficient. However, this road did not run in parallel with environmental concerns. Despite the attention given to these, they were not an important driving force of the reform process. The composition of the parliamentary working group's staff indicated the dominance of economic concerns. The Environmental Protection Agency (EPA) was not represented which it would have been if environmental concerns had been the central motives of the reform. Furthermore, both the EPA and the Agricultural Board, which also had responsibility for agri-environmental matters, questioned the positive environmental effects of the reform. They even feared that environmental problems might worsen (Proposition 1989/90: 146, bilagedel, pp. 64, 85, 88). The dominance of economic concerns was emphasised by the minister of agriculture who stated in the cabinet bill that 'the reform of the food policy is ... an important measure to keep down inflation' (Proposition 1989/90: 146, p. 36, my translation). However, he also pointed out that environmental protection needed more attention than the parliamentary working group had given it (ibid., p. 50). Since this emphasis on environmental protection came relatively late in the process, environmental concerns legitimised the reform rather than driving it.

Reform proposals recommending second order reforms of the agricultural policy

In the mid-1980s, the costs of agricultural policy began to mobilise actors outside the agricultural policy network. These were especially concerned with the economic welfare losses caused by the high price policy.

The first analysis to raise fundamental questions about the agricultural policy was undertaken by a professor at the Swedish Agricultural University, an official from the Swedish Industrial Association and a professor at Lund University. It was published in 1984 - six months ahead of the publication of the food policy commission's report. At that time, the fundamental questions which the analysis raised had the attention of neither the Commission on Food Policy nor the public. The questions raised by Bolin and his associates were: do we need an agricultural policy, and can the objectives of agricultural policy be fulfilled by the market instead of public regulations? They also questioned the commission's ability to deal with the fundamental problems of agricultural policy. The authors focused upon surplus production, high food prices, the financial problems of young farmers and the lack of transparency of the agricultural policy. To overcome these problems, the authors argued that deregulation was the best solution (Bolin et al., 1984). Their book triggered a public debate on the agricultural policy in which leading newspapers took an active part (Rabinowicz, 1992, p. 289; interview, Board of Agriculture, 1994a).

After Bolin and his associates had set a new agenda for the agricultural policy debate, public authorities started asking similar types of questions. In 1986, the minister of agriculture set up a one-man commission to investigate the factors influencing food prices. The investigator, Lars Hillbom, was a former official at the Confederation of Trade Unions and had represented consumers in the Consumer Delegation. The two secretaries were officials at the National Price and Cartel Office (*Statens Pris- och Kartelnämnd*) which had a central position in the investigation since it prepared 11 of the commission's 12 sub-reports. Farmers were weakly represented in the expert committee attached to the commission. They had only one of 19 seats (Micheletti, 1990, p. 106; SOU 1987: 44, pp. 3-13). Even in Sweden, it was unusual to have a person loyal to consumer interests heading a commission investigating agricultural policy.

While the commission was in progress, the *Riksdag* debated the foundation of war readiness. Up until this point, the agricultural policy stated that food production capabilities had to be large enough to withstand a three year embargo. New national security investigations suggested that it was unrealistic that could be totally cut off from international trade for so long. Instead, the Department of Defence assessed that a total embargo would last no longer than few months (Proposition 1986/87: 95, p. 37). This was a major setback for farmers because war readiness legitimised a large excess capacity in the means of production and, consequently, considerable overproduction (Feldt, 1991, p. 345).[13] The new point of departure for the planning of wartime agricultural production influenced the commission's evaluation of the agricultural policy. Taking this development into account, the commission concluded that food production should be decreased and that competition should be increased by allowing more imports. Direct subsidies to farmers should replace price subsidies. Increased stockpiling of food would ensure the fulfilment of the war readiness objective (SOU 1987: 44, p. 9-11, 138). The report did not lead to policy changes, but it indicated that farmers were beginning to lose control over the future policy process.

The consumers joined the camp of reformers which was beginning to form. Compared to other countries, consumer groups in Sweden are well-organised. They have formed the Consumer Delegation (*Konsumentdelegationen*) which negotiated directly with farmers in the annual agricultural price negotiations.[14] One of its member associations, the Central Confederation of Professional Employees (*TCO*), published its recommendations for a new agricultural policy same year as the one-man commission. Like the one-man commission, the Confederation recommended that the high price policy be replaced with a low price policy in which farmers were subsidised through acreage payments. The Confederation argued that reduced prices required other support forms of which direct payments was the most obvious. Leaving farmers to free market forces was not considered (TCO, 1987, p. 52-3). Both the one-man commission and the consumer report recommended a second order reform of the agricultural policy. The state should continue to take responsibility for imbalances in the agricultural sector but by employing new policy instruments.

The 1990 agricultural policy reform

In 1984, the Department of Finance began to examine the problems of agricultural policy, aiming at fundamental policy change. Kjell-Olof Feldt, the minister of finance, wanted total deregulation including a subsequent abolition of import protection (Feldt, 1991, p. 346). The agricultural investigations were conducted by the Expert Group for Studies in Public Economy (*Expertgruppen för studier i offentlig ekonomi*) and by the Agency for Administrative Development (*Statskontoret*) (Finansdepartementet, 1987, p. 3; Micheletti, 1990, p. 109). In particular, two reports attracted attention among policy makers and became central in the policy process. Both reports were published in autumn of the 1988.

The first report was published by the National Defence Research Institute (*Försvarets Forskningsanstalt*). It concluded that the existing policy only had a limited contribution to war readiness because modern agriculture required stable supplies of fertilisers, machinery and electricity. Ensuring stable input supplies was not at all guaranteed by the high price policy (FOA, 1988, p. 98). More importantly, the report concluded that in peacetime import levies and price regulation could only ensure some types of wartime food supplies. For instance, if certain parts of Southern Sweden were occupied by enemy forces, food shortages would occur in the rest of the country (ibid., p. 36). The major conclusion of the report was that deregulation of the agricultural policy would be no obstacle to the fulfilment of the war readiness objective (ibid., p. 102). Therefore, the report indirectly recommended a third order policy reform; that is, the state should no longer take economic responsibility for imbalances in the agricultural sector. Instead, farmers should be subject to the conditions of the market place, like other industries. The Federation of Swedish Farmers, LRF, strongly opposed the report's conclusions, arguing that in the case of crisis or war, a liberalised agricultural sector could not guarantee food supplies. It also criticised the analysis for lacking biological insight and it questioned the calculations. Not surprisingly, consumers supported the conclusion of the Defence Research Institute report; however, they were anxious that a non-regulated market would lead to instability in the food market. The Agricultural Marketing Board (*Statens Jordbruksnämnd*), which administers the price policy, adopted a neutral position on the conclusions of the report but questioned some of its premises and calculations. Likewise, the Agricultural Board (*Lantbruksstyrelsen*) which administers the structural policy was careful not to pick sides; however, it agreed with the Defence Research Institute's conclusion that a deregulated agricultural sector could fulfil the war readiness objective (Jordbruksdepartementet, 1990). The cautious line of the two agricultural state boards left farmers in a bad situation. They seemed to be without strong allies within the state apparatus.

The Department of Finance published the second report. It concentrated on the economic costs of agricultural policy. In 1986, the costs were 13.7-14.6 billion Swedish *kronor* which was nearly double that of 1979.[15] Further, the report concluded that price subsidies caused inefficient allocation of economic resources because high commodity prices in agriculture directed resources away from sectors in which they could be more efficiently applied. Consumers were the losers of agricultural policy as they paid the bill for agricultural support (Finansdepartement, 1988). To overcome these problems, the report recommended that the import protection be lowered and the domestic market deregulated. To compensate farmers for part of their income losses, the report suggested the introduction of a direct payment scheme in which support was decoupled from production volume and farm size. Not all farmers would be entitled to direct subsidies and there would be an upper limit on support. Full-time farmers who were most in need seemed to have priority (ibid., p. 99). Whereas the Defence Research Institute's report had indirectly recommended a third order reform, the report of the Department of Finance directly recommended a change in policy which came close to a third order reform. However, the report did not suggest the total abolition of import protection which the minister of finance seemed to favour. This would probably have been too drastic and thus not politically feasible. It would hit Swedish agriculture very hard. World market prices would then apply in Sweden, and since they were kept down by large export subsidies, Swedish agriculture would only have a limited chance of survival.

LRF, the Farmers' Federation, did not like this report. It put forward the same type of critique as it did on the Defence Research Institute's report; that is, criticising the analysis and the conclusions. Instead of deregulation, farmers wanted to retain the existing system. They argued that a direct support scheme was not attractive for them. Subsidies should relate to performance and not just to being a farmer. In other words, they did not want subsidies to look like social benefits. The Consumer Delegation welcomed the report and supported its conclusions and recommendations but wanted further analysis of direct payment systems to be undertaken before it took up a position. The Agricultural Marketing Board and the Agricultural Board recognised that the existing policy had a low degree of achievement of aims. To decrease the costs of the agricultural policy, the Marketing Board argued that it was necessary to alter the objectives of the policy in a way which would lead to lower financial support. However, the Board argued that reductions in production through price cuts could be implemented within the existing policy, if the policy makers changed the production goals. The introduction of direct income support schemes to lessen farmers' income losses resulting from lower prices did not have the support of the Board. It argued that such a policy instrument was difficult to administer. The Agricultural Board opposed a reduction in import protection. It supported the idea of some reduction in internal market regulation but did not seem to favour a total elimination of the market regulation. Nor did it support direct payment schemes (Jordbruksdepartementet, 1990a). Clearly, the two agricultural state boards did not support a third order reform. They were not yet ready for more than a considerable first order reform. However, they refrained from allying with the farmers in a strong coalition in defence of the status quo.

In the beginning of the reform process in 1987, the minister of finance and the Department of Finance strengthened their position. The cabinet appointed two political officials from the Department of Finance to positions in the Department of Agriculture where they were assigned to head the reform process. The two, Michael Sohlman and Gunnar Wetterberg, were selected with care. Sohlmann who was appointed state secretary was known as 'one of Kjell-Olof Feldt's boys' (*Land* no. 41, 9 October 1987, my translation) and was the former Budget Director in the Department of Finance. While in that position, he had prepared an agricultural reform paper which was too radical for some members of the cabinet (Feldt, 1991, p. 347). Gunnar Wetterberg was the author of the Department of Finance report recommending agricultural deregulation.

In December 1988, the minister of agriculture set up a parliamentary working group to evaluate the agricultural policy and draft a reform proposal. The group consisted of Members of parliament from all political parties represented in the Riksdag. Neither farmers nor consumers were full members of the working group, but they had advisory status (Jorbruksdepartementet, 1989). Although LRF was not a full member of the parliamentary working group, the association considered itself as such (*Land* no. 44, 4 November 1988). Sohlman was appointed as chairman of the group and Per Molander, who wrote the Defence Research Institute's report, became its secretary (Jorbruksdepartementet, 1989). Contrary to previous state commission investigations, the secretaries of the working group were people from outside the agricultural policy network (OECD, 1995, p. 49; interview, Board of Agriculture, 1994a). Micheletti (1990, pp. 184-5) observes that the status of the group was unique. Firstly, it was not a state commission and, secondly, interest associations were given a minor role. The working group allowed the LRF representatives to participate in only some meetings. Until the middle of August 1989, they had, by and large, been excluded from the preparation of the report (*Land* no. 33, 18 August 1989). Furthermore, they were not allowed to submit minority statements (LRF, 1990a, p. 1). The marginalisation of farmer representatives was not the only sign of the exclusion of agricultural interests. Although the director generals of the Agricultural Marketing Board and the Agricultural Board took part in the work, they seemed to have only limited influence (*Land* no. 12, 23 March 1990).

In October 1989, the working group submitted a report stating that the 1985 policy had not fulfilled its objectives. On the whole, it supported the conclusions of the reports of the Defence Research Institute and the Department of Finance, both of which had accounted for the problems caused by the agricultural policy. New investigations were also carried out. The working group concluded that there was no income problem in agriculture; the problems of agriculture were equal to those of other industries This was somewhat unexpected because farmers regarded this as an almost chronic problem. The report was also critical of the use of price subsidies to raise farmers' income. Attempts to raise their income had failed because the price support had driven up the price of farm land which, in turn, had decreased the incomes of young farmers. Nor did the price subsidies fulfil regional objectives, the group concluded. In addition, the fulfilment of the policy's environmental aim was also disappointing. The system of price subsidies was blamed for being one reason for this. Lower prices would decrease nitrate run-offs (Jordbruksdepartementet, 1989). As for the costs of exporting surplus production, the state's spending turned out to be almost three times as high as estimated (ibid., p. 92). In 1985, the state's share was estimated at 600 million Swedish *kronor* until 1990 but it turned out to be 1.7 billion *kronor*.

The working group recommended considerable alterations in the objectives of the agricultural policy. Ensuring food supplies in peacetime, during trade embargoes and in wartime in a way which took the use of natural resources and the environment into consideration was the main objective of the 1985 policy. The sub-goals were to ensure that consumers could buy food of good quality at fair prices and to ensure that farmers' living standard was equal to that of comparable groups in Swedish society (JoU 1983/84: 20 p. 7; JoU 1984/85: 33). The working group seriously threatened the special status of agriculture, arguing that agricultural 'production should be subject to the same conditions as other industries' (Jordbruksdepartementet, 1989, p. 165, my translation). For this reason, the working group wanted to change the policy objectives. The war readiness objective should no longer guide peacetime food production to the same extent as previously since the peacetime production capability did not say anything about what it would be in wartime. This implied that specific production goals should not be formulated. To ensure crisis and wartime food supplies, stockpiling of food was to be extended. With regard to the sub-goals of the 1985 policy, the working group recommended that the environmental and the consumer objectives should be retained (ibid., p. 167). As for the income goal, the group concluded: 'What is basically demanded are those goods and services agriculture can contribute to; and if the goals in these regards (war readiness etc.) are fulfilled, then payments for labour and capital have been sufficient' (ibid., p. 167, my translation). In other words, farmers should derive their incomes not from state subsidies but from the market. There would no longer be a safety net under farmers' income and it would be very difficult for farmers to argue for high levels of peacetime production to guarantee wartime food supplies.

Having assessed the consequences of deregulation, the parliamentary working group concluded that the objectives could be fulfilled through a reformed policy in which domestic market regulations were abolished (ibid., pp. 238-40). In a choice between lowering prices within the existing system or a fundamental reform, the members of the working group chose the latter. It recommended that:

> ... the existing system with border protection and internal regulation of prices and production is replaced with border protection without internal market regulations. The border protection is to be supplemented with goal-directed publicly funded measures in order to consider the demand for war readiness, to protect valuable elements of arable land and to strengthen regional policy (ibid., p. 251, my translation).

The new policy involved a third order reform which was to be fully implemented by 1995. A positive effect emphasised by the group was that deregulation would decrease the use of fertilisers and chemicals; in that, it would lead to lower prices which, in turn, would lead to extensification, decreasing nitrate pollution by 20 per cent by the mid-1990s (ibid., p. 221). To conserve landscapes in areas which would be hurt by afforestation as a result of the deregulation, the working group recommended that the state increase the conservation funds considerably (ibid., pp. 252, 259, 265). All the members of the parliamentary working group supported deregulation; however, the Members of parliament from the Centre Party and the Conservative Party wanted a more precise formulation of the objectives. The former also wanted a precise goal for farmers' income (ibid., pp. 269-320).

Not surprisingly, LRF was not pleased with the report. Firstly, the association was displeased that it had been marginalised in the parliamentary working group (LRF, 1990a, p. 1). Secondly, LRF criticised the analyses of the report for being weak and incomplete and stated they should not form the basis of the *Riksdag's* decision to abolish the internal market regulations (ibid., pp. 2, 4). However, LRF could accept a gradual alterations of the exiting policy. Against this background, the association presented its reform proposal. The objectives of the 1985 policy should remain unchanged. Likewise, the present level of import protection should be maintained and import levies, commodity levies and levies on means of production should continue to fund export subsidies. Moreover, LRF argued that the level of state subsidies should be maintained or even raised. Maintaining the system of export subsidies and import protection would allow a considerable simplification of the internal market regulations. In the cereals sector, LRF recommended that the state introduced a long term switch-over programme to limit cereals surpluses by stimulating farmers to use their land for other purposes, for instance growing alternative crops (LRF, 1990a, 1990b, pp. 10-17). Farmers held the state financially responsible for the transition (Öberg, 1994, p. 206). Although farmers claimed that they were prepared to accept changes in the existing policy, the alterations they suggested would basically have involved no more than a first order reform. There were no indications of a general shift in policy instruments in LRF's reform paper, nor was the basic policy principle, the principle of state responsibility, to be changed. Since surplus exports could still be subsidised, there would, in practice, have to be a domestic minimum price on which the export subsidy could be based (LRF, 1990b, p. 15). Therefore, domestic minimum prices could be indirectly set through the export subsidies and thus kept at a high level. In conclusion, the state should, to a large extent, continue to protect farmers from market forces.

Consumers were much more pleased with the report. They supported the proposal to alter the policy objectives and to abolish the internal market regulation. This would have the effect of curbing food price increases and improving the market signals from consumers to producers. However, the Consumer Delegation would have liked to go even further in agricultural deregulation by lowering the import protection in order to introduce more competition into the food market (Proposition 1989/90: 146, bilagedel, pp. 18, 43, 71). The Agricultural Board which had previously been rather sceptical of the reform proposal of the Department of Finance now became much more positive to fundamental policy change, supporting the argument that farmers should be subject to the same conditions which applied to other industries. In general, the Board favoured a policy reform leading to a more market-orientated price setting on farm products (ibid., pp. 37, 42). Similarly, the Marketing Board had also become more positive towards fundamental reform. It agreed with the parliamentary working group that there were important reasons to limit the regulations which, in turn, would give agriculture conditions that were more like those of other industries. Therefore, the Marketing Board supported the abolition of the internal market regulation (ibid., pp. 37, 43). Both boards called for better measures to facilitate the transition (ibid., pp. 157, 159). The policy positions of the two boards clearly showed that farmers had not succeeded in obtaining the support of agricultural state actors. Farmers had, therefore, become isolated in the reform process.

The parliamentary working group argued that a reformed policy would benefit the environment by diminishing nitrate pollution (Jordbruksdepartementet, 1989, p. 221). However, both the Agricultural Board and the Environmental Protection Agency questioned this conclusion. They feared that further specialisation and concentration of agricultural production might be the result of the reform. This would lead to increased nitrate run-offs in some areas. The Swedish Society for Nature Protection expressed similar concerns (Proposition 1989/90: 146, bilagedel, pp. 85, 88, 99).

Having agreed that the abolition of the internal market regulation was the overall aim of the forthcoming reform, the political parties had taken a major step forward in changing the agricultural policy. What was left for the cabinet was to draft a detailed proposal. By and large, the cabinet based its bill on the recommendations of the parliamentary working group. The cabinet shared the working group's view that agriculture should not have conditions which were different from those of other industries (Proposition 1989/90: 146, p. 50). Compared to the recommendations of the working group, the bill was a little more sympathetic to agricultural interests. In the cereals, dairy and meat sectors, the transition measures were more advantageous to farmers. For instance, the cabinet changed the transition period in the cereals sector from three to four years, the dairy sector was to be gradually deregulated over a three-year period instead of a rapid change and the temporary direct payments in the arable sector was supplemented by land conversion schemes (ibid., pp. 57, 72, 165-6). Compared to the recommendations of the parliamentary working group, the environmental goal received more emphasis in the cabinet bill (ibid., p. 49).

Before the cabinet presented its bill to the *Riksdag*, the minister of agriculture had negotiations with the Farmers' Federation but they broke down after few meetings (*Land* no. 15, 12 April 1990). Disagreement over who should bear the costs of the reform seemed to have caused this. LRF claimed that farmers were to pay a large share of the transition although the state contributed with temporary subsidy schemes (*Land* no. 17, 27 April 1990). Nevertheless, in the dairy sector, the cabinet did agree with LRF on the reform (JoU 1989/90: 25, p. 62).

The *Rikdag's* Standing Committee on Agriculture consulted the three parties of the agricultural network dealing with the price policy, i.e. LRF, the Consumer Delegation and the Agricultural Marketing Board. Further, it met with delegations from producer interest associations and received letters from public authorities, organised interests and enterprises (JoU 1989/90: 25, p. 30). The committee accepted the new policy objectives but did not accept some elements of the bill. There was agreement that the state took more economic responsibility for the transition. This was a concession to farmers. It provided more economic resources for the transition schemes and extended the transition period by one year in the dairy and the meat sectors while reduced the transition period in the arable sector by one year. Giving in to agricultural pressure, the *Riksdag* raised the funds available for conversion schemes in the arable sector, prolonged the period in which the schemes would be in force and broadened the range of purposes for which support would be given. Contrary to the intentions of the cabinet bill, the *Riksdag* did not deregulate the oilseed sector (JoU 1989/90: 25, pp. 1-2, 46-64; OECD, 1995, pp. 51-8). All political parties in the *Riksdag*, except for the Green Party, could then support the policy reform (Riksdagen, 1990; Rabinowicz, 1992, p. 287). Despite these alterations, '[t]he changes made by the parliament in the reform package did not change the ultimate objective of the reform package which was to achieve a complete internal deregulation within a five-year period' (OECD, 1995, p. 52). Therefore, the reform remained a third order reform.

Changes in the administrative system implementing agricultural policy constrained farmers' opportunities for turning the implementation process in a direction which would lessen the impact of the reform (Rabinowicz, 1992, p. 284). The implementation remained within the agricultural sector but policy makers carried out a comprehensive reorganisation. The Agricultural Marketing Board amalgamated with the Agricultural Board and became the Swedish Board of Agriculture (*Statens Jordbruksverk*). Farmers were not pleased with this development because, in their view, the new board took over the attitudes of the Marketing Board. In general, these attitudes were more sympathetic to consumer interests than those of the Agricultural Board (interview, de Woul of LRF, 1994). Before the amalgamation, the governing board of the Marketing Board included both farmer and consumer representatives. In the Agricultural Board, farm workers counterbalanced farmers (SOU 1990: 87, pp. 10, 23). In the new Board of Agriculture, farmers, consumers and farm workers lost their seats. The governing board now consists of persons representing the 'public interest'. These persons are appointed on the basis of their personal qualifications rather than on their affiliations to certain interests associations. The previous influence of the governing board has been reduced (ibid., p. 65; interview, Board of Agriculture, 1994a). This development clearly curbed the influence of farmers.

The abolition of the semi-public regulatory boards which carried out the day-to-day administration of the market regulation was a major setback for farmers' opportunities to influence the implementation process. Farmer representatives had dominated these boards as they possessed detailed, technical information and knowledge which put them in an advantageous position compared to the boards' state and consumer representatives (Micheletti, 1991, pp. 95, 125). The Board of Agriculture took over the day-to-day administration of agricultural policy (Proposition 1989/90: 146, p. 187). Although the reform removed many regulatory functions, the decision not to do away with the import protection left some regulation which demanded advice from the interests concerned. Therefore, the new Board of Agriculture established market advising committees for a number of commodities. These committees have both consumer and farmer representation. Besides the market committees, there are some advisory committees dealing with other agricultural matters (SOU 1990: 87, p. 78; interview, Board of Agriculture, 1994a).

Contrary to what one would expect, the reactions of the Farmers' Federation, LRF, were relatively mild (*Land* no. 22, 1 June 1990; no. 23, 8. June 1990). The LRF president, Bo Dockered, said that compared to what the parliamentary working group had presented, the reform enacted had come closer to LRF's position (*Land* no. 23, 8. June 1990). Öberg (1994, p. 216) concludes that historically the Swedish Federation of Farmers has valued the advantages of close relations with the state more highly than the disadvantages. Seen from the Federation's point of view, the major advantage of having close relations with the state was that it was clearly the best way of gaining influence. The disadvantage was that LRF had moderate demands. In the 1990 reform process, the original aim of LRF was to maintain the existing policy, but realising that this was politically impossible, LRF moderated its demands, hoping for a deal with the state (ibid., pp. 212-13). Even though LRF did not reach a general compromise with the state, it chose not to mobilise opposition after the *Riksdag* had decided to reform the policy. Instead, LRF tried to get the best out of the situation and to preserve its close relationship with the state. To do so, the leaders of LRF put a great deal of effort into persuading the rank-and-file members to accept the new agricultural policy (ibid., pp. 209-11).

Farmers were not the only ones who were critical of the reform. The initiator of the reform, the minister of finance, was also sceptical but for other reasons. He preferred a tougher solution in which the state's share of the transition costs was much lower and he disliked the fact that the *Riksdag* removed important elements of the deregulation. He also had to accept that the import protection was upheld. In the early stages of the reform process, he had advocated for a reduction (Feldt, 1991, pp. 346-8).

To sum up, the Swedish agricultural policy reform was a third order policy reform. The farmers were unable to mobilise sufficient political support to prevent radical reform mainly because the Agricultural Marketing Board and the Agricultural Board did not ally with them. Instead, the two boards placed themselves between the farmers and the reformers.

On the very first day of the new agricultural policy, Sweden applied for membership of the European Community (interview, Board of Agriculture, 1994a). The application for EC membership meant that the Swedes never fully implemented the reform. Instead, they began to adjust their agricultural policy to the Common Agricultural Policy. For this reason, the government reintroduced the market intervention system which included guaranteed minimum prices and export subsidies. To match the Mac Sharry reform, it also introduced direct compensation payments (OECD, 1995, pp. 95-6). Thus, the Swedish experiment in agricultural policy reform was never fully realised. But one must not forget that Swedish policy makers did make the tough decision to reform the agricultural policy - a decision which only the New Zealanders had dared to make. Contrary to what one might think, there are no indications in the sources of evidence which has been analysed here that anticipation of future EC membership had any influence upon the reform. The Swedish government made the decision to apply for EC membership in the autumn 1990, several months after the reform was enacted. The fact that the reform was not adapted to the Common Agricultural Policy is yet another indication that there is no causal link between the anticipated EC membership and the Swedish agricultural policy reform.

Conclusion: comparing the agricultural policy reforms and the positions of agricultural authorities

The agricultural policy reforms in the EC and Sweden were very different. In the EC, the reform was much more moderate than was it in Sweden. It involved a change of policy instrument in the arable sector and an alteration of instrument settings in the animal sector. The EC continued to protect farmers from free market forces; there was no intention of leaving farmers to the mercy of the market place. By switching from price support to direct payments, the Common Agricultural Policy in the arable sector changed from a high price to a low price policy. In other words, the CAP reform involved a second order reform in the arable sector and a first order reform in the animal sector. Swedish policy makers enacted a more radical policy reform. They changed the policy objectives, the instruments and the state's responsibility for the well-being of farmers was considerably reduced. The principle of state responsibility was no longer in force; thus, the Swedish reform was a third order policy reform.

EC and Swedish agricultural authorities chose different policy positions in the reform process. The EC agriculture commissioner and the Agricultural Directorate protected farmers against attacks aimed at changing the basic policy principle of the CAP. This principle puts forward the idea that the EC should take responsibility for imbalances in the agricultural sector. In Sweden, at the beginning of the reform process, the Agricultural Board and the Agricultural Marketing Board were sceptical of a third order policy reform. Later they even supported the idea, but not wholeheartedly. Farmers also reacted differently to the reform proposals. EC farmers would not accept more than modest first order reforms while Swedish farmers would accept more comprehensive first order reforms. Because Swedish farmers had only weak support from agricultural authorities, they were forced to come up with a proposal which the other participants could take seriously. This was the only way to gain influence. On the contrary, EC farmers could pursue a confrontational strategy because the agriculture commissioner and the Agricultural Directorate protected their basic interests.

In both the EC and Sweden, there were intense and acrimonious debates on the reform proposals. There was a major difference in these debates. In the EC, third order reform was not even on the agenda. The conflict between the Commission and farmers was not over whether the EC should or should not subsidise agriculture, but over how it should be subsidised. The agriculture commissioner and the Agricultural Directorate controlled the reform process and therefore they totally excluded the question of third order reform from the agenda. In Sweden, the conflict between farmers and the state was much more fundamental as it involved a debate on whether or not the state should subsidise farmers. The reformers gained control of the agenda and, as a result, they put the issue of third order reform on to it. Unlike EC agricultural authorities, Swedish agricultural authorities were unable to place themselves at the centre of the reform process and, consequently, they had only limited influence.

Notes

1. There is some confusion about the terms European Community and European Union. Policy making within the first of the three pillars of the European Union is within the European Community. The first pillar deals with most policy areas, e.g. agricultural policy, the single market project and competition pol-

icy. The second pillar concerns foreign policy and the third one legal affairs. Since agricultural policy making is within the European Community, I use that term.

2. For instance, the United States applies a low price agricultural policy.

3. The GATT (General Agreement on Tariffs and Trade) is a multi-lateral agreement on international trade. The aim of the GATT is to further international trade and economic development. Agreements on reduction in tariffs and abolition of trade barriers are the means to achieve these aims (Udenrigsministeriet, 1986).

4. Later, the Commission estimated that the German reunification only accounted for one-seventh of the increase (*AE* no. 1433, 28 March 1991, p. P/1).

5. The beef situation worsened later on. In April 1991, the Commission estimated that the beef spendings would increase by 100 per cent in 1991 (*AE* no. 1437, 26 April 1991, p. P/2).

6. It consists of fourteen countries which are Argentina, Australia, Brazil, Canada, Chile, Colombia, Hungary, Indonesia, Malaysia, New Zealand, the Philippines, Singapore, Thailand and Uruguay.

7. The EC had another view of the world trade problems. It believed that the deficiency payments of the US low price agricultural policy had the same effects on world trade as the EC export subsidies . The only difference, according to the EC, was that US export subsidies were indirect. However, these arguments were not valid for Australia and New Zealand since their agricultural support was minimal compared to that of the EC. In 1988, the proportion of the agricultural production value which would be put down to direct and indirect agricultural state subsidies (the percentage PSE) was 46 in the EC, 8 in Australia and 7 in New Zealand. Since 1988, the percentage PSE has decreased to 3 per cent in New Zealand in 1993 (OECD, 1994, p. 107).

8. The power of the Agricultural Directorate within the Commission is indicated by the fact that it 'is the only sectoral branch of the Commission "placed in the driver's seat for the actual conduct of the [GATT] negotiations"' (Keeler, 1996, p. 137).

9. Provided that the target price and the world market price are at the same level. This price level was estimated at 100 ECU/t. If world market prices fail to reach that level, the EC cereals prices will be 90 ECU/t which is the EC intervention price (the guaranteed minimum price). In that case, the price reduction would be 42 per cent (*AE* no. 1448, 112 July 1991, p. E/2).

10. The examination of the reform process in the Council is based mainly on Landbrugsministeriet, 1991a, Landbrugsministeriet, 1991b, Landbrugsministeriet, 1991c, Landbrugsministeriet, 1991d, Landbrugsministeriet, 1991e, Landbrugsministeriet, 1992, Landbrugsministeriet, 1992a, Landbrugsministeriet, 1992b, Landbrugsministeriet, 1992c, Landbrugsministeriet, 1992d, Landbrugsministeriet, 1992e, Landbrugsministeriet, 1992f, Landbrugsministeriet, 1992g.

11. See chapter 5 for a review of its environmental recommendations.

12. For similar reasons, Feldt tried to persuade the minister of agriculture, Svante Lundkvist, to work for deregulation of the agricultural policy in 1982. The Agriculture Minister turned down the proposal, saying that he would rather resign than challenge the power of farmers (Feldt, 1991, pp. 113-4).

13. The war preparedness objective was not the only safeguard farmers had against change but it was the most important one. The Swedes also emphasised the regional employment effects derived from the agricultural policy. Another objective of the policy was to maintain economically sustainable farms in some areas to prevent these from afforestation (Jordbruksdepartementet, 1989, pp. 135, 152). To some extent, these two objectives were also safeguards against fundamental agricultural policy reforms.

14. In chapter 7, I deal with the consumer's role in Swedish agricultural policy making. In chapter 8, it is argued that change in the Centre Party's and in the Social Democrat's loyalties paved the way for consumer influence. In the late 1950s, the Centre Party transformed from a party primarily representing farmers' interests into a party which also represented the middle class. In the same period, the Social Democrats downgraded their concern for the small farmers and began to emphasise consumer interests.

15. According to the PSE measure which sums up all direct and indirect agricultural producer subsidies.

7 Cohesion of agricultural policy networks

Introduction

In chapter 5, Danish and the Swedish nitrate policies were compared. Chapter 6 undertook a similar comparison of agricultural policy reforms in the European Community (EC) and in Sweden. Both comparisons also focused on how the political actors, particularly the agricultural authorities, positioned themselves in the policy processes. In nitrate policy making, Danish farmers were more successful in transferring the principle of state responsibility to nitrate policy making than their Swedish colleagues who had to accept that the polluter pays principle came to play an important role. As a result, the Danish nitrate policy is a low cost policy whereas the Swedish is a high cost policy.

Farmer interests remained privileged in the European Community because the principle of state responsibility continued to be the basis of the Common Agricultural Policy (CAP) after the reform in 1992. Swedish farmers failed to defend this principle and, as a result, market principles to a large extent became the foundation of the 1990 agricultural policy. While the EC adopted a second order policy reform in the arable sector and a first order reform in the animal sector, Sweden enacted a third order reform.

Agricultural authorities took up different positions in the four cases. In Sweden, the agricultural authorities refrained from supporting farmers. Consequently, farmers were, so to speak, left on their own, unable to prevent fundamental changes from taking place. Danish agricultural authorities were very sympathetic to farmers' interests in the national nitrate policy process. EC farmers achieved considerable support from the agriculture commissioner and the Agricultural Directorate in the 1992 policy reform. For this reason, Danish and EC farmers were better off than Swedish farmers.

With the help of the policy network model developed in chapter 2, one can provide a meso-level explanation of why Swedish farmers could not mobilise the support of the agricultural authorities to the same extent as farmers in Denmark and in the EC. The lack of sufficient political support was a major reason why Swedish farmers, in contrast to Danish and EC farmers, failed to transfer the principle of state responsibility to nitrate policy making and failed to defend it successfully in agricultural policy making. The theoretical network model suggests that variation in the cohesion of agricultural policy networks helps to explain policy differences. A cohesive network binds members together when they face pressure for change from outside. Under such conditions, agricultural authorities are likely to support the interests of farmers. Because farmers know they can rely on such support, they often choose a confrontational strategy towards outsiders. Danish farmers chose such a strategy. A less cohesive network produces a different actor grouping. The lack of cohesion makes it difficult for the network's members to form strong coalitions against outsiders. As a result, farmers cannot rely on the support of agricultural authorities and, thus, they do not apply confrontational strategies towards outsiders. Instead, they pursue cooperative strategies, believing that this is the best way to handle the situation. Swedish farmers preferred cooperative strategies. Cohesion occurs in policy communities whereas it is absent in issue networks. A policy community is cohesive because it has only a few members who represent a narrow range of interests. The members are well integrated in the policy process and the degree of institutionalisation is high. In contrast, an issue network has a large number of members who represent a wide range of interests and are only partially integrated in the policy process. Additionally, there is only a low degree of institutionalisation.

The purpose of this chapter is to examine the cohesion of agricultural policy networks in Denmark, Sweden and the EC to test whether or not the policy network model, developed in chapter 2, provides a meso-level explanation of the variation in policy choices. For the model to be externally valid, the empirical examination must show that the degree of cohesion of the agricultural policy networks in Denmark and the EC is higher than in Sweden. If the analysis finds such a pattern, then I provide further evidence for the model which argues that the structures of policy networks have important influence upon the policy preferences of agricultural authorities and state actors in general. In addition, if the above empirical pattern is found, then the findings also support the proposition arguing that network structures influence the strategies of farmers' associations.

We cannot establish the degree of cohesion directly; to examine it, we have to apply indicators. It has been argued that a policy community (one extreme on the network continuum) is the most cohesive type of network and an issue network (the other extreme) is the least cohesive. Establishing the empirical networks' position on the continuum developed in chapter 2 indicates whether there are differences in cohesion. The theoretical network model is valid if the Danish and the EC agricultural policy networks come closer to a policy community than the Swedish.

The network continuum has three analytical dimensions. Firstly, it emphasises the importance of the network's membership and the range of interests which participants represent. Secondly, the way in which the members are integrated in the policy process affects the degree of cohesion. Finally, the continuum emphasises institutionalisation. This refers to the extent to which there is a consensus on policy principles and on procedures to approach policy problems. Policy principles define what should be the content of public policy and procedures to approach policy problems describe what members regard as problems and how they should be examined.

While establishing the membership of a policy network is relatively straightforward, analysing integration and institutionalisation require greater efforts. It is difficult to establish directly whether bargaining/negotiation or consultation prevails in a network. Usually, one cannot gain access to data which can indicate the extent to which one or the other mode of interaction is predominant. Therefore, one has to use a research method which examines the frequency of formal and informal interaction within the network and which analyses the organisational arrangements within which interaction takes place. When contacts are frequent and organisational arrangements to facilitate interaction are well-developed, there are reasonable grounds for concluding that bargaining predominates. This means that the network has a high degree of integration. In contrast, if the pattern of interaction within a network is unstable and there are only loose organisational arrangements to facilitate interaction, then it is unlikely that bargaining can take place. Since consultation is likely to be the predominant mode of interaction within such a policy network, we can conclude that its degree of integration is low.

Full analysis of the institutionalisation of a network requires vast amounts of high quality data. Since this research project does not have the resources and the time needed, it will only examine one aspect of institutionalisation. The aspect focused on is the extent to which there has been a consensus on the basic principle of agricultural policy - the principle of state responsibility - *within* the agricultural policy network. We can establish this by using an indicator which analyses the extent to which the objectives of agricultural policy have remained stable, and whether the members of the network, over time, questioned the agricultural policy objectives. If the objectives have been stable and if there have been no serious disputes *within* the network, then there are reasonable grounds to conclude that there is a strong consensus on the principle of state responsibility, indicating that the network has a high degree of institutionalisation. Conversely, changing objectives and conflicts over agricultural policy objectives suggest that a consensus on the principle of state responsibility is non-existent within the network. This indicates that the network's degree of institutionalisation is low.[1] We can only establish whether there is a consensus on policy principles within a network through a study over time. Limiting the examination to a certain point in time or a short period suggests invalid results (see Sabatier, 1993, p. 16; Smith, 1993, p. 8). For example, what looks like a consensus may be only a temporary compromise which will be questioned by one or more members from time to time.

Denmark[2]

The Danish agricultural policy network has been, and still is, very cohesive. The Ministry of Agriculture and the agricultural organisations have been the core members of the policy network for more than six decades. A high degree of integration has characterised the network and the main objectives of the national Danish agricultural policy in the pre-EC period remained stable. After the Danish entry into the EC in 1973, the members of the network have pursued the same objectives as earlier but within the context of the Common Agricultural Policy.

Between the beginning of 1932 and March 1933, the relationship between state and agriculture in Denmark changed fundamentally. These changes laid down the framework for agricultural politics in the decades to follow. From being an industry relying on free market forces, agriculture went into close cooperation with the state. State intervention came to mark the period from the beginning of the 1930s until today. Before the depression in the 1930s, farmers had a strong belief in a liberal free trade (Just, 1994, pp. 38-9; 1995, p. 13). The state had only a minimal role in the agricultural sector except during the First World War when the state intervened for foreign policy and distributive reasons (Just and Omholt, 1984, p. 45).

The situation in the export markets in the beginning of the 1930s was the reason for the breakdown of the Danish liberal free trade agricultural policy. Germany, Belgium and Great Britain restricted Danish food imports by imposing tariffs and quotas. Since these countries were the three main export markets for Danish food, the restrictions caused enormous difficulties for the Danes. The belief in the market forces as the best solution to agricultural problems could not be upheld and, in consequence, the agricultural associations had to give up their reluctance to state intervention. They even asked the state to intervene (Just, 1992, pp. 90, 107-9, 121).

The restrictions imposed upon Danish exports forced farmers and the state into close cooperation because the state had to authorise central regulation of the exportation of food. Export boards were set up for various commodities in order to allocate quotas between exporters, to ensure that differentiated tariffs did not distribute export earnings unfairly, to find new markets and to ensure that production was reduced without creating chaos. Further, boards were set up to regulate the production through the application of individual quotas, on pigs in particular (Just, 1992, ch. 3-4). In the domestic market, levies were imposed on butter and later also on milk for human consumption to maintain minimum prices (Tracy, 1989, p. 210; Just and Omholt, 1984, p. 51). The export boards were set up under the Ministry of Agriculture, but it was only represented on three of the ten boards.[3] Farmers' organisations became central actors in the policy network which developed. Each board consisted of representatives from the cooperative association involved in the production and export of the commodity concerned. Representatives from commercial food processors and exporters also became members of the boards. In all boards, the agricultural interests had the majority but, what is more important, the association representing farmer cooperatives and the Council of Agriculture carried out the day-to-day administration (Just, 1992, pp. 96-121, 517; 1994, p. 39; 1995, p. 13).[4] Surprisingly, the commercial processors and exporters had no objections to this (Just, 1992, p. 117).

There is no doubt that the administration of the export regulation provided the agricultural associations with an important political power base. They had a monopoly on detailed and highly valued information on the export and domestic markets. Therefore, the government had to consult them if it intended to introduce new legislation. Furthermore, the minister of agriculture could not, under the Agricultural Export Act, take any initiative in agricultural export policy unless the Council of Agriculture recommended it. In other words, the agricultural associations had significant control over agricultural policy, leaving the commercial processors and exporters in a relatively weak position in the policy network. In the early 1930s, the agrarian protest movement, LS (*Landbrugernes Sammenslutning*), was the largest farmers' union, but despite this, it was almost excluded from the agricultural network even though it had some support in the *Folketing* (Danish Parliament), in particular from the Conservative Party. LS had only two representatives on one of the ten boards. Consumer and labour interests were also almost completely excluded from the network. The only case of labour representation was on the Butter Export Board on which there was one representative from the Dairymen's Union (Just, 1992, ch. 3-4, pp. 515-6). Until 1950, the agricultural associations strengthened their representation on all the boards. On the five most important boards it increased from 59 per cent of the seats in 1933 to 73 per cent in 1949 (Just, 1994, p. 518).

Some would perhaps question whether the Ministry of Agriculture was a member of the agricultural policy network since it only had seats on three export boards. Even though the ministry was weakly represented on the boards, it did indeed have a central position in the network due to the possession of important political resources. It had legal and political expertise on which the other members depended. Without the support of the ministry, the agricultural associations and the commercial processors and exporters had severe difficulties persuading the cabinet to adopt certain measures. Indeed, the ministry did refuse to follow the recommendations of the agricultural associations on several occasions (Just, 1992, pp. 100, 104, 227).

In 1950, the government abolished the state export boards but the agricultural associations replaced them by similar private boards on which the interests of agriculture and the commercial food processors and exporters had seats. Agricultural interests continued to dominate the boards with 73 per cent of the seats (Just, 1992, p. 518). The contacts between agriculture and the state remained close (Just and Olesen, 1995, p. 134). In the late 1950s, agricultural exports again ran into difficulties. Prices decreased while costs increased (Buksti 1974, pp. 206-8). The creation of the European Community in 1957 caused major problems since Denmark as a non-member had no access to the EC market (Betænkning no. 561, 1970, p. 7; Tracy, 1989, p. 271). In 1957, the farmers' associations asked for state intervention in order to cope with agri-economic problems (Betænkning no. 561, 1970, p. 105; Buksti, 1974, p. 60).

In general, the model to handle the economic problems of agriculture was similar to that of the 1930s, with the exception that the smallholders' and the farmers' unions obtained a much stronger position in the network compared to the farmers' cooperatives (Buksti, 1974, pp. 67-70). Boards consisting of agricultural associations and commercial food processors and exporters governed the implementation of agricultural policy, and the agricultural associations carried out day-to-day administration. As in the 1930s, the agricultural associations had the majority in the governing boards. This was not the only indication that the commercial food processors and exporters had limited influence within the network. In 1969, the agriculture minister set up a committee to scrutinise the agricultural policy. Surprisingly, the commercial food producers and exporters had no seats on the committee, thus indicating that they did not have a central position within the network (Betænkning no. 561, 1970, p. 5). The agricultural policy of the 1960s consisted of various cereals regimes, direct subsidies to farmers, tax reliefs and domestic market regulations (Buksti, 1974, pp. 255-63; Just, 1994, p. 40; Betænkning no. 561, 1970, ch. 6). The purpose of the market regulations was to maintain domestic market prices above world market prices. During the 1960s, agricultural subsidies for agriculture increased from 4 per cent of farmers' net factor income in 1960 to 33 percent in 1968 (Just and Omholt, 1984, p. 65). Although the Ministry of Agriculture did not participate directly in the implementation of the agricultural policy, it still had a central position in the network. It had the political and legal expertise upon which the organised interests depended. It had the power to refuse recommendations from the two other members of the network, even when they agreed on these recommendations (see Buksti, 1974, pp. 82, 263 for examples).

Denmark's entry into the European Community in 1973 changed the organisation of agricultural policy making. From being formally and highly integrated into the policy process the organised interests were formally only given an advisory role; however, informally the agricultural associations continued to occupy a central position within the policy network. The Ministry of Agriculture increased its political and administrative resources, thus strengthening its position in the network significantly. Acting as the Danish negotiator in Brussels provided the ministry with a new major resource. Insight into the EC policy process was a highly valued resource among the other members of the network. The ministry's new role entailed that it increased its staff considerably and, therefore, strengthened its political and administrative resources. Further, the implementation of agricultural policy was fundamentally reorganised. While the ministry had earlier delegated the day-to-day policy administration to the agricultural associations, it now carried out the implementation in its newly established Directorate for the Market Regimes (later the EC Directorate). Originally, the agricultural associations and the Ministry of Agriculture intended to apply the existing model for agricultural policy administration in a modified form. There was strong opposition to this from the other ministries and from the cabinet. This persuaded the Ministry of Agriculture to administer the policy itself (Christensen, 1978, pp. 91-2, 108-12). These changes resulted in a weakening of the information monopoly of the agricultural associations (Buksti, 1974, p. 263; 1980, p. 288).

Formally, the degree of integration within the network significantly decreased because of the Danish EC membership, but, informally, change was only marginal. To advise the agriculture minister before the meetings in the EC Council of Ministers and to advise on the administration of the Common Agricultural Policy in Denmark, the government set up an advisory committee, consisting of 17 members

(Administrationsdepartementet 1987, bilag 2). They represented the Ministry of Agriculture (two members), the Ministry of Industry (two members), the Confederation of Danish Industries (two members), the Danish Chamber of Commerce (two members), the Economic Council of the Labour Movement (one member), the Consumer Council (one member) and the agricultural associations (seven members) (Lov no. 595, 1972). The agricultural members represent the farmers' cooperatives and the Farmers' and the Smallholders' unions. Although the committee has no formal powers, it is an important forum for the agricultural policy process. Firstly, the representatives express their official positions at the meetings. Secondly, the committee sets up working parties, usually consisting of civil servants from the ministry or its agencies and officials from the agricultural associations. Usually, policy decisions are made in these working parties, meaning that the influence of the other members of the advisory committee is very limited (Pedersen, 1981, p. 81; Landboforeningerne, 1991, p. 25; *Landsbladet* 9 October 1992, interviews, Ministry of Agriculture, 1992, DDB, 1992, DFU, 1992).

The informal contacts between the Ministry of Agriculture and the agricultural associations are still well developed, in particular in the pigmeat and dairy sectors (Landbrugsministeriet, 1977, p. 89, Christensen, 1981, p. 197; interviews, Ministry of Agriculture, 1992, 1993b, DDB, 1992, Council of Agriculture 1992, DFU, 1992). This type of contact takes place in telephone conversations and in more or less regular sectoral commodity meetings (Nedergaard et al., 1993, p. 49). Exchange of working papers, minutes, memoranda etc. are important means of informal integration. Informal interactions between the ministry and the agricultural associations include also the ministry's civil servants at the Danish Permanent Representation in Brussels. They often meet with officials based at the Brussels office of the Danish Council of Agriculture (Bregnsbo and Sidenius, 1993, p. 193; interview, DDB, 1992). Meetings between the minister of agriculture and the presidents of the agricultural associations take place several times during the year; however, since the early 1980s, the meetings have become less frequent (interview, DFU, 1992). The leaders of the agricultural associations also meet occasionally with politicians from political parties represented in the *Folketing*. Christian Sørensen who has been the president of the Smallholders' Union for 20 years said in an interview that these contacts were more frequent than people realised (interview, Sørensen, the Smallholders' Union, 1993).

The European Community's introduction of milk quotas in 1984 brought the Danish agricultural associations back into their former role in agricultural policy administration. Milk quotas were controversial among Danish farmers and were very difficult for the Ministry of Agriculture to administer because it did not have access to databases containing information on individual milk deliveries, including information on the percentage of fat in the milk delivered. For these reasons, the ministry suggested that the Dairy Board administered them (Lindrup, 1986, p. 57). The Dairy Board set up the Milk Committee to administer the quota scheme. By acting as an appeal body and by supervision, the ministry ensured a certain control over the Milk Committee's administration of the quotas (Danske Mejeriers Mælkeudvalg, 1992, pp. 10-14; Administrationsdepartementet, 1987, bilag 3). Administering the individual quotas requires detailed information on milk deliveries. The Dairy Board was the only organisation possessing the information to implement the scheme successfully (Lindrup, 1986, pp. 82, 92). As a result, the Ministry of Agriculture was dependent upon the Dairy Board. On the other hand, the Dairy Board depended upon the ministry to grant it legal authority to force dairy farmers to comply with the Milk Committee's decisions on individual quotas.

The non-agricultural interests represented in the advisory committee have not been able to position themselves centrally in the agricultural policy network. This is, for example, indicated by the fact that they do not attend the committee meetings on a regular basis (Pedersen, 1981, pp. 79-80; interview, Ministry of Agriculture, 1992). EC membership has resulted in a loss of influence for the commercial food processors and exporters. Before Denmark's entry into the EC, their major concern was access to export markets, a concern they shared with the agricultural associations. By opening the European market for Danish food exports, EC membership solved many problems of the commercial food processors and exporters. Thus, their major incentive to participate in agricultural policy making disappeared. They did not participate in the two committees set up in the 1970s and 1980s to discuss the future agricultural policy (Betænkning no. 795, 1977, p. 6; Betænkning no. 993, 1983, pp. 7-8).[5] Since the commercial food processors and exporters seem to share many interests with agriculture (Daugbjerg, 1992, p. 46), they may have felt that the agricultural representatives who dominated both committees took care of their interests. Believing that trade takes place at all price levels, commercial traders do not perceive the annual price setting negotiations as important (Pedersen, 1981, p. 79) because high prices do not imply that international competitors take over market shares. The EC food market is protected from the world market.

The consumers play a minor role in the network (interviews, Ministry of Agriculture, 1992, DFU, 1992). They claim that the agricultural associations dominate the meetings in the ministerial advisory committee, and that the ministry and the agricultural associations make deals before the meetings (interview, Danish Consumer Council, 1993). Moreover, they hold that the ministry is mainly concerned with agricultural interests. Therefore, the Consumer Council has always been sceptical of agricultural ministers (*Jyllands-Posten* 16 May 1996). The Labour Movement also plays a minor role (see Landbrugsministeriet, 1987, pp. 79-81). Since it favours a high level of production due to employment concerns (interview, Danish Consumer Council, 1993), its interests, to a large extent, coincide with those of agriculture.

The ministry's informal contacts with the non-agricultural members of the advisory committee are almost non-existent. Their views are not valued in the policy process unless they are in agreement with agricultural interests (interview, Ministry of Agriculture, 1993b).[6] Recently, the Ministry of the Environment seems to have achieved slightly more influence in the agricultural policy network. Since Denmark entered the EC, it has been a member of the special committee for agriculture which coordinates the Danish ministries' positions in the EC agricultural policy. The Ministry of Agriculture dominates the meetings and since the meetings take place relatively late in the decision making process, it is difficult for the other ministries to exercise influence (Telephone conversation, Ministry of Agriculture 1993). Therefore, the Ministry of the Environment now enters the policy process in earlier phases (interview, Ministry of the Environment, 1993). The Ministry of Finance also has seats on the special committee for agriculture. It plays only a minor role in Danish agricultural policy making. Since finance ministers are mainly interested in the net contribution of the Common Agricultural Policy (Nedergaard, 1988, p. 58), the Danish minister of finance has had no reason to counterbalance agricultural interests. Denmark benefits from the Common Agricultural Policy; thus, agricultural and budgetary interests coincide.

To establish the degree of institutionalisation in the Danish agricultural network, I examine the extent to which there has been a consensus on the basic principle of agricultural policy making: the principle of state responsibility. Since we cannot reveal the principle directly, an indicator which examines how stable the agricultural policy objectives have been over time is applied. Stable policy objectives indicate that there is a consensus on the principle of state responsibility which, in turn, indicates that the degree of institutionalisation is high (see the introduction to this chapter). Danish agricultural policy has, however, never developed formal objectives; they have remained informal.

The dominating objective of the Danish agricultural policy until the entry to the EC in 1973 was to maintain the international competitiveness of agricultural production in order to uphold a high level of export earnings and a high level of employment in the agricultural sector, especially in the food processing industry. Maintaining farmers' incomes has been another important objective. This objective has always been linked to the objective of international competitiveness. Farmers' incomes should, to a large extent, depend upon the ability to compete in the exports markets. EC membership did not change anything in this respect. Although the EC ensured that prices and outlets were stable, Danish farmers still had considerable interest in maintaining their international competitiveness. Selling in the market has always been more profitable than selling to EC intervention stocks.

Until 1973, none of the members of the network seriously criticised the objectives of the agricultural policy. The objectives did not conflict with the interests of the commercial processors and exporters. In fact, they benefited from the upholding of a competitive Danish agricultural sector because they were as vulnerable to foreign import restrictions as farmers' cooperatives. Therefore, they had no incentive to oppose the income objective. Nor did high domestic food prices and the provision state subsidies to farmers hurt their interests. Indeed, commercial food processors and exporters also benefited from state subsidies applied to promote Danish food abroad.

In the late 1960s, political actors outside the agricultural policy network criticised the use of state subsidies to support farmers' incomes. The most serious criticism came from the Economic Council which argued that these subsidies could be applied much more efficiently in other sectors of the Danish economy (Det Økonomiske

Råds Formandskab, 1968).[7] This criticism forced the minister of agriculture to set up a committee to scrutinise the agricultural policy. Representatives from the agricultural associations and the Ministry of Agriculture together had a comfortable majority. The committee concluded that besides maintaining farmers incomes, the agricultural subsidy schemes had the purpose of upholding agricultural production until Denmark entered the European Community. EC membership would solve the export problems of Danish agriculture (Betænkning no. 561, 1970, pp. 18, 151). To sum up, from the early 1930s until the early 1970s, a consensus on the principle of state responsibility characterised the agricultural policy network.

Participation in the Common Agricultural Policy moved Danish agricultural policy making away from the problems of market access and, consequently, the commercial food processors and exporters became less central; as a result, farmer interests came to dominate. EC membership strengthened the consensus on the principle of state responsibility within the agricultural network because the Community took over the responsibility for imbalances in the agricultural sector. This meant that the costs of agricultural policy were passed on from the Danish state to the consumers and taxpayers in the other EC member states. Since 1973, Denmark has derived big financial benefits from the Common Agricultural Policy (Nedergaard et al., 1993, pp. 134-5). Therefore, maintaining, and even increasing, farmers' incomes came to coincide even more with the national interest. Since the Danish state did not have to pay the costs of agricultural policy any longer, the consensus between the two core members of the agricultural network became further strengthened. There was, and still is, a high degree of consensus between the Ministry of Agriculture and the farmers that the competitiveness of Danish agriculture should be maintained and that farmers should earn a decent income. Both parties agree that the EC should ensure that these objectives are achieved. As Buksti (1983) points out: 'Motivated by basic national interests, of both a specifically agricultural and a general economic nature, the Danes favour a 'European' solution to financing problems of the CAP.' Clearly, such a solution is in the interest of Denmark because it obtains a considerable economic net benefit from the EC membership (Buksti, 1983, pp. 267-8; Tracy, 1993, pp. 251-2). In other words, there are close links between agricultural and national interests (Buksti, 1980, pp. 288-9; 1983, pp. 268-9, 277-80). In the 1970s and early 1980s, there was agreement within the network that it was in the national interest to expand the agricultural production for employment and balance of payment reasons (Betænkning no. 795, 1977, p. 142; Betænkning no. 993, 1983, p. 13). A reading of the Ministry of Agriculture's publications leaves no doubt that it perceives these economic roles of agriculture as very important. For instance, the report on sustainable agriculture which is published by the ministry starts by saying that 'agriculture is an important job creating and foreign currency earning sector in the Danish economy' (Landbrugsministeriet, 1991, p. 11, my translation). For these reasons, the ministry believes that 'agriculture will continue ..., as an exporting industry, to be a considerable asset for Danish economy (Landbrugsministeriet, 1986, my translation).

Members on the periphery of the agricultural policy network, except for the

Consumer Council, do, to a large extent, accept the policy principles of the network. The consumers are critical of the principle of state responsibility as they believe that the market should set food prices; however, consumers are not against farmers receiving direct income support (interview, Danish Consumer Council, 1993). Thus, they do not totally oppose the principle of state responsibility, but are against the way it is applied. Despite its opposition to the core members of the network, the Consumer Council has not been able to mobilise serious opposition to the principle.

The period from the early 1970s until the mid-1990s has also been characterised by a consensus on the principle of state responsibility within the Danish agricultural policy network. The network comes close to the ideal type of a policy community because it has few core members representing a narrow range of interests; because the degree of integration is high and because there is a consensus on the policy principles. We cannot characterise the network as a genuine policy community since it has a periphery in which a wider range of interests is represented.

Sweden

Like Denmark, Sweden experienced a turning point in agricultural politics in the early 1930s. Agriculture became a highly regulated industry with close relations with the state. However, the Swedish agricultural policy network did not become as stable as the Danish network. Firstly, consumers achieved a central position within the network, and, secondly, since the early 1960s, agricultural policy has been characterised by changing objectives. Therefore, the Swedish network is clearly not as cohesive as the Danish.

During the late 1920s, cereals prices decreased (Thullberg, 1977, p. 191). In 1930, this development triggered state intervention into the cereals sector to stabilise the market. This included guaranteed minimum prices and an obligation for all mills to use at least 60 per cent Swedish cereals in production.[8] The administrative model of the cereals regime laid down the principles for future agricultural policy administration. Administration of the regulatory system was delegated to a semi-public body, the Swedish Grain Association which was governed by a board consisting of state representatives and millers (Seyler, 1983, p. 173). Cereals were not the only problem facing policy makers. The dairy sector also had severe problems. Even though Swedish agricultural exports played a numerically minor role in the agricultural economy, decreasing world market prices had serious consequences for farmers. A dramatic fall in butter prices[9] caused falling domestic milk prices and, consequently, domestic competition increased (Thullberg, 1977, pp. 191-3; Hellström, 1976, p. 338; Rothstein, 1992, p. 110; Seyler, 1983, p. 170).[10]

Falling world market prices were not the only factor causing increasing domestic competition. New transportation technology in the market for milk for human consumption was another important reason for the agricultural crisis. During the 1920s, the use of lorries for milk transportation disturbed the order in the milk market. Before lorries were used, dairy farmers in densely populated areas received a higher price for milk for human consumption than farmers who produced milk for processing. The latter lived in remote, sparsely populated areas. The lorry made it possible for them to enter the market for milk for human consumption. As the supply of this increased, prices decreased, thus threatening the profitability of the traditional producers. When the butter prices on the export markets declined, the situation in the market for milk for human consumption became even more chaotic. Former producers of butter now switched to milk for human consumption, thus further decreasing milk prices (Hellström, 1976, pp. 334-6; Rothstein, 1992, p. 111, Seyler; 1983, p. 170).

In 1932, the *Riksdag* (the Swedish parliament) set up a dairy market regime. In respect of policy administration, the devolution of public authority went even further in the dairy sector than in the cereals sector because the Swedish Dairies' Association administered the regime. In order to level out milk prices, it had the right to collect levies from all dairy farmers. The Department of Agriculture set up a board to oversee the administration of the regime (Hellström, 1976, pp. 577-8). From 1933 to 1934, the government continued to intervene in agriculture. It set up market regimes for pigmeat, beef, potatoes and eggs and introduced a levy on wheat in the cereals regime (Hellström, 1976, pp. 572, 586, 591, 597). Similar to the cereals regime, semi-public bodies, overseen by market boards, administered the market regimes (SOU 1977: 17, p. 238; Hellström, 1976, ch. 10). In 1935, the market boards were merged into one: the National Agricultural Marketing Board (*Statens Jordbruksnämnd*). Agricultural interests dominated its governing board until 1948 when consumer representatives were given seats (Rothstein, 1992, pp. 237, 247).

In the decades following the Second World War, consumer influence gradually increased in agricultural policy making (Elvander, 1969; Micheletti, 1990; Steen, 1988; Rothstein, 1992, ch. 12). In 1952, consumer representatives decided to influence agricultural policy more systematically. The Cooperative Union and Wholesale Society, the Confederation of Swedish Trade Unions and the Central Confederation of Professional Employees formed the Consumer Committee. However, the committee came to play only a limited role (Elvander, 1969, p. 241). It was later abolished and replaced by the Consumer Delegation (*Konsumentdelegationen*), which was attached to the National Agricultural Marketing Board in 1963 as an advisory body (e.g. SOU 1984: 86, p. 176, Elvander, 1969, p. 242; Micheletti, 1990, pp. 94-5). Originally, the Consumer Delegation consisted of representatives from the trade unions, the Cooperative Union and Wholesale Society but in 1977, wholesale trading companies and business associations not belonging to the farmers' movement became members (Michelleti, 1990, pp. 95, 127).

The Consumer Delegation had the right to participate in price negotiations between the farmers and the state, but had no clear status (Steen, 1988, p. 216). Although its status has never been codified by law, the Consumer Delegation later became an equal negotiating partner. Since the mid-1970s, the Swedish state has also provided it with financial resources (Micheletti, 1990, pp. 96-7, 133; SOU 1984: 86, p. 176; Proposition 1984/85: 166, p. 78; JoU 1984/85: 33, p. 66). Micheletti (1990, pp. 94-7, 133) and Steen (1988, pp. 214-18) conclude that, since the early 1970s, the Consumer Delegation has been relatively influential in agricultural policy making.

However, on the one hand, the presence of consumers in the network did not result in lower levels of agricultural subsidies in Sweden than in the EC. On the other hand, consumer representation may have prevented the level of subsidies from increasing to the Norwegian level, one of the highest levels in the world.[11] The most important influence of the consumers in Swedish agricultural politics is that they prevented the agricultural network from developing a high degree of cohesion (Daugbjerg, 1996a; 1997; 1998). Compared to other Western countries, the strong position of Swedish consumers in agricultural policy making is unique. Two circumstances in the late 1950s decreased the structural power of farmers in the *Riksdag* and thus explain why consumer interests gained strength. Firstly, after having supported smallholder interests since the early 1920s, the Swedish social democrats started to emphasise consumer interests. This shift in social loyalties motivated the social democratic minister of agriculture to initiate the establishment of the Consumer Delegation in 1962 (Steen, 1988, p. 214). Secondly, farmers also lost support within the Agrarian Party. It changed its name to the Centre Party, symbolising that the party had become a party representing the interests of white-collar workers as well as the farmers' interests (ibid., pp. 339-40).[12]

It would not have been possible for consumers to gain access to the agricultural policy network's core without a change in the views of the agricultural state bureaucracies. Since the early 1960s, the Department of Agriculture has become less of an agricultural client department and more of a mediator between conflicting interests. Firstly, the Department of Agriculture became more orientated towards consumer interests. In fact, it suggested the establishment of the Consumer Delegation in 1963 (Elvander, 1969, p. 242). The government bill which established the Delegation stated that it came naturally that the state set up such a body (Steen, 1988, p. 214). In the 1970s, the Department of Agriculture became even more consumer-orientated (Micheletti, 1990, p. 101). One reason for this may be that, compared to other countries, government departments in Sweden are small and do not deal with day-to-day policy administration. Therefore, they may be more responsive to the policy preferences of the minister, the state secretaries and other political appointees. So, in the final analysis, the department's change in position may be the result of changes within the political parties, particularly within the Social Democratic Party. Secondly, the department became responsible for environmental policy making in 1967. A special unit (*Miljövårdsberedningen*) was set up to advise and to provide information on environmental policy making. The unit consisted of experts, cabinet members and senior civil servants. Until 1987 when the Department of the Environment was established, the minister of agriculture chaired the meetings of the unit (Lundqvist, 1971, pp. 148-9; Lundqvist, 1996, pp. 270, 273). That the Swedish Department of Agriculture broadened its concerns becomes evident when one compares the number of civil servants employed in its agricultural division and in its environmental division. As early as 1983, the agricultural division employed seven civil servants, whereas twice as many worked in the environmental division (*Statskalender*, 1983).[13] The Agricultural Marketing Board has also changed from a client bureaucracy into an interest mediator. In 1962, it opposed attempts to give consumers and farmers equal status in the annual price negotiations. Like the farmers, the Marketing Board preferred the existing system of bilateral negotiations (Micheletti, 1990, p. 133). During the decades to follow, it developed into a mediator. For example, in the 1990 agricultural reform process, it tried to balance between consumer interests favouring deregulation and agricultural interests seeking only adjustments of the old policy (see chapter 6). An LRF official whom I interviewed said the Marketing Board was the farmers' economic counterpart. He did not think the Board had been 'working for the interests of agriculture, but had been inspired by consumer interests and the public interest' (interview, de Woul of LRF, 1994, my translation).

Some actors have not gained access to the core of the agricultural network. The core members of the network have not valued these actors' resources. However, they are not totally excluded from the policy process because the Swedish system of public consultations provides them with an opportunity to express their views. For instance, the Farm Workers' Association and the Association of Workers in the Food Processing Industry are on the periphery of the network (Bolin et al. 1984, pp. 76-80). Their employment concerns seem to carry little weight in the policy making process. Competing farmers' unions, like the smallholders, also belong to the outer circle (SOU, 1984: 86, p. 420). Nor has the Environmental Protection Agency been involved in agricultural policy making (interview, EPA, 1994). Though the Department of Finance has an interest in limiting the costs of agricultural policy, it does not belong to the network's core. Analysing the relations among Swedish government departments, Petersson (1989, pp. 61-2) demonstrates that the contacts[14] between the Department of Finance and the Department of Agriculture are half as frequent and intensive as the contacts between the Department of Agriculture and the Ministry of the Environment. Even the Department of Foreign Affairs has closer contacts with the Department of Agriculture. Preventing the agricultural regulatory system from spreading to other sectors rather than counterbalancing agricultural interests seems to have been the major concern of the Department of Finance until the mid-1980s (Feldt, 1991, p. 345).

The Farmers' Federation, LRF, and the Consumer Delegation are formally integrated into the annual price setting negotiations. In the first stage of these negotiations, the Agricultural Marketing Board (or a committee appointed by the government), LRF and the Consumer Delegation try to reach a framework agreement for the price setting within a certain period. Usually, the state representatives have a modest role in this phase. If a compromise is reached, the cabinet presents a bill to the *Riksdag*. Since the framework agreement is not binding, the cabinet and the *Riksdag* can make alterations, but, usually, the *Riksdag* has approved the bill without major changes. In the second stage of the price setting, LRF and the Consumer Delegation enter into negotiations in which they agree on a detailed price proposal. The Agricultural Marketing Board then assesses the proposal and presents it to the cabinet which makes the final decision (SOU 1984: 86, pp. 76-8, 418-20; Steen, 1988, pp. 204-07; Bolin et al., 1984, pp. 73-4; Micheletti, 1990, p. 132).

The informal integration of the farmers and consumers into the agricultural policy process has been just as comprehensive as the formal. Petersson (1989, pp. 120-1) shows that in the 1980s, the Department of Agriculture had relatively close contacts with both LRF and consumer representatives: 50 per cent of the department's officials had contacts with LRF at least sometimes during the year, and 42-47 per cent had contacts with groups representing consumers.[15] However, the figures do not tell us much about the intensity of the contacts. The term 'at least sometimes during the year' includes great variation in the frequency of contacts. Also, the quality of the contacts may vary considerably. Despite this reservation, the figures indicate that, informally, consumers and farmers are about equally integrated into the agricultural policy network.

It has been argued above that the extent to which agricultural policy objectives are stable indicates a policy network's degree of institutionalisation. There were several changes in the agricultural policy objectives from the early 1930s until the 1990 reform. This indicates that a consensus on the state's role in agriculture did not develop among the core participants of the network. Until 1947, agricultural policy gradually came to take more social concerns into consideration. Thullberg (1974, p. 166) argues that in 1940, agricultural policy was seen as part of Sweden's social policy and Vail, Hasund and Drake (1994, p. 67) points out that '[i]ncome maintenance was the policy's prime objective, though it was not yet expressed in the form of an explicit income target.' In the light of the unstable political situation in Europe, war readiness, i.e. ensuring food supplies, also became an increasingly important policy objective in the late 1930s.

The 1947 agricultural policy was the first attempt to coordinate the state's intervention into agriculture (SOU, 1977: 17, p. 109). The social democrats, who had considerable influence on the 1947 policy, 'favored a policy that was based more on social than on economic considerations' (Micheletti, 1990, p. 54). The social element of the policy was reflected by the main objective of the agricultural policy which was to ensure that farmers' average income was equal to that of industrial workers (Steen, 1988, pp. 63, 74). Ensuring that Sweden remained self-sufficient in food supplies during crises and war times also became an important objective (Micheletti, 1990, p. 53).

Until the early 1960s, consumers had little influence on the agricultural policy. After the Consumer Delegation was formed in the early 1960s, the consensus on policy objectives within the network disappeared. The 1967 policy reflected the growing consumer influence in the agricultural policy network. In Sweden, trade unions represented consumers' interests. They pursued influence on the agricultural policy because '[f]ood prices and the costs of regulation [were] seen as directly related to the economic situations of their own membership' (Michelleti, 1990, p. 73). To ensure reasonable food prices, consumer representatives favoured a policy which was based more on market forces. Consumers achieved influence on the 1967 policy which stated that Swedish agriculture should, to a larger extent, be subject to the conditions that applied for business and industry. The income objective lost precedence and it became less precise. Instead, the main objective became a production goal stating that Sweden should produce 80 per cent of its food demand. In addition, agricultural production should increase the efficiency through structural development (SOU 1977: 17, pp. 110-13; Steen, 1988, pp. 76-9; Micheletti, 1990, pp. 91-3).

In 1977, the pendulum swung back towards more social and non-economic concerns in agricultural policy. The income goal was again given a higher priority and the production target was raised so that the degree of self-sufficiency was to be 100 per cent. New concerns which emphasised regional development and employment, environmental protection and consumers' access to high quality foodstuffs at fair prices were also introduced (SOU 1984: 86, pp. 75-6; Steen, 1988, pp. 79-80; Micheletti, 1990, pp. 101-03). In the mid-1980s, there was a limited move back to more market forces in agriculture because policy makers downgraded the income objective. Furthermore, farmers became more responsible for the costs of exporting surplus production. The state would pay only 40 per cent of the export subsidies whereas farmers were responsible for the remaining 60 per cent (Proposition 1984/85: 166; JoU 1984/85: 33). This indicated that there was a move towards more market forces in agriculture. Ensuring food supplies in peacetime, during trade embargoes and in wartime in a way which took the use of natural resources and the environment into consideration was the main objective of the 1985 policy. The sub-goals were to ensure that consumers could buy quality food at fair prices and to ensure that farmers' living standard was equal to that of socially equivalent groups (JoU 1983/84: 20, p. 7; JoU 1984/85: 33).

Unstable policy objectives are associated with a lack of consensus on the state's role in agriculture. In particular, the frequent changes in the status of the income objective suggest that there was no consensus on the principle of state responsibility within the network. In other words, there was a conflict within the network over the extent to which the state should protect farmers from the market forces. The entrance of consumers into the core of the agricultural policy network was a major reason that a consensus on the principle of state responsibility did not develop. Therefore, the degree of institutionalisation was low.

One could argue that the changing objectives of Swedish agricultural policy may not be associated with a lack of consensus on the principle of state responsibility. If both farmers and consumers agreed to alter the objectives, then there would be a consensus. However, there was no consensus between farmers and consumers. Had they agreed on the principle, farmers would have accepted the consumers' membership of the network because it would have legitimised the content of agricultural policy even further. The reluctance of the Farmers' Federation, LRF, to accept the Consumer Delegation as an equal negotiating partner is a strong indication that the agricultural policy network lacks consensus on the principle of state responsibility. In 1984, the Farmers' Federation argued that the Consumer Delegation had no role to play as the farmers' negotiating partner in the agricultural policy process since it had nothing to offer farmers. Consequently, LRF could not make mutual agreements with the Consumer Delegation. LRF regarded the state as its only true negotiating partner; however, LRF stated it would accept to negotiate with the Consumer Delegation if the state devolved it authority to represent state interests (SOU 1984: 86, pp. 525-6; see also *Land* no. 40, 6 October 1989). In the farmers' opinion, the Consumer Delegation did not represent its constituency. As the LRF president, Bo Dockered, said in 1989: 'the Consumer Delegation has had its day and should be abolished as it is not rooted among consumers' (*Land* no. 24, 16 June 1989, my translation).

In conclusion, the Swedish agricultural policy network was far from being a policy community since consumers belonged to its core. Therefore, the range of interests represented in the network was not narrow. Furthermore, the degree of institutionalisation was low.

The European Community

Which institutions of the European Community should be included in the analysis of the agricultural network? As Kassim (1994) correctly points out, it is difficult to delineate EC policy networks. This book is mainly concerned with the EC agricultural authorities' position in the 1992 reform process. Hence, I examine the Agricultural Directorate General's and the agriculture commissioner's relations with other political actors in the process in which the policy proposals were prepared. Since organised interests rarely approach the Council as an institution (Nugent, 1991, p. 232; Butt Philip, 1985, p. 42) and since the Council does not negotiate directly with interest groups, it is not likely to become involved in the early phases of the policy process. Therefore, in analysing the preparation of reform proposals, it is reasonable not to include the Council of Agriculture Ministers in the agricultural policy network. However, what goes on within the Agriculture Council has a crucial impact on the final policy choices.

The EC agricultural policy network is relatively cohesive. Farmers, the agriculture commissioner and the Agricultural Directorate (DG-VI) dominate the network. However, farmers have diverging interests, resulting in an inability to act in unison. Despite this, they are well integrated in the agricultural policy process. Since the formation of the Common Agricultural Policy (CAP) in 1958, there has been no change in the policy's objectives.

The Treaty of Rome was signed by Germany, France, Italy, the Netherlands, Belgium and Luxembourg in 1957. Article 38-47 of the treaty obliged the Community to set up a common agricultural policy based on certain principles. The policy was to be fully implemented by the end of 12-year transition period (De europæiske Fællesskaber, 1987, pp. 253-61; Fearne, 1991, pp. 25-6). In accordance with the Treaty, the six member states held a conference in Stresa in Italy which laid down the specific principles of the Common Agricultural Policy. These formative phases of the policy had a crucial impact years ahead. From the outset of the Common Agricultural Policy, agricultural interests were privileged. Representatives from farmers' unions and, with the exception of Italy and Luxembourg, the food industry were included in the national delegations attending the Stresa conference (Neville-Rolfe, 1984, p. 195). The consumer interest was poorly represented but the interests of farmers and the food industry were, to a very limited extent, counterbalanced. Agricultural workers had a seat in the Dutch delegation (Robinson, 1961, p. 133). Despite this, consumer interests 'did not attract a great deal of attention either in the working party reports or, with two exceptions, in the ministers' speeches' (Neville-Rolfe, 1984, p. 198). Few months after the conference, the farmers' unions in the EC member states formed COPA (*Comité des Organisations des Producteurs Agricoles*) and the agricultural cooperatives formed COGECA to represent their interests at EC level. Sicco Mansholt, the first agriculture commissioner, encouraged the formation of these two associations (Averyt, 1977, p. 54; Phillips, 1990, pp. 51, 83). In order to limit consultation with farmers to one association and to aggregate European level agricultural expertise, the Commission established the principle of consulting only Euro-groups (Burkhardt-Reich and Schumann, 1983, p. 332).

During the early 1960s, the market organisation of the Common Agricultural Policy was established. In 1962, the Council of Ministers reached agreement on the cereals regime which then became the model for several other market regimes. By July 1968, the market regimes of the CAP were fully implemented (Tracy, 1989, pp. 255-6, 265). The Commission consulted both COPA and farm workers' unions on the establishment of the market organisation. It had several meetings with COPA to 'explain its ideas and to obtain the reactions of organised agriculture in the six countries' (Robinson, 1961, p. 134). The farm workers' unions were 'mainly interested in structural and social problems rather than the details of market policy' and, therefore, they pointed out the need for a social policy for agriculture (ibid., p. 135). Since the interests of farmers and farm workers, to a large extent, coincided, giving farm workers access to the policy process did not provide a countervailing power to farmers. Consumers had direct interests in the market organisation of the Common Agricultural Policy, but until 1973 they were completely excluded from the agricultural policy process. So, agricultural interests had a privileged position which they used to influence the CAP.

Until the late 1970s, COPA maintained its central position in EC agricultural policy making. The other core member of the agricultural policy network was the Agricultural Directorate (DG VI). DG VI succeeded in gaining control over agricultural policy making although the Commission was a collective body; that is, a commissioner must achieve the approval of the whole Commission before s/he can put forward a formal proposal. In the 1970s, DG VI controlled the agricultural policy process, enabling it to prepare policy proposals without the interference of the other directorate generals. An attempt to involve the directorate generals for external affairs, economic affairs and the budget in the agricultural policy process failed (Phillips, 1990, p. 51). DG VI had no interest in consulting others because this might have led to increased pressure on the CAP. Ensuring the maintenance of the agricultural policy had high priority for DG VI officials. The Common Agricultural Policy was their whole *raison d'être* (Smith, 1990a, p. 155; Nedergaard, 1994, p. 95; 1995, pp. 130-1; Keeler, 1996, p. 137). Consequently, most of those involved in the policy process were status quo minded (Gardner, 1987). In spite of the attempt to open the agricultural policy process to other directorate generals, there was a fairly high degree of consensus on agricultural policy within the Commission. Within EC institutions, there was a strong belief that the Common Agricultural Policy was the driving force of European integration. Indeed, there was a certain pride that the European Community had succeeded in establishing a common policy (Atkin, 1993, pp. 56-7; see also Grant 1995, p. 15; Keeler, 1996, p. 136).

COPA enjoyed a privileged position within the agricultural policy network throughout the 1970s. The introduction of the so-called objective method in the annual price review in 1973 helped to ensure its central position. The objective method calculated 'the percentage increase in guaranteed prices necessary to keep incomes of 'modern' farms in line with non-farm incomes' (Phillips, 1990, p. 54). In order to provide input to the calculation of prices, considerable bureaucratic re-sources were needed to obtain detailed information on the economic situation of both farmers and non-farmers in the whole Community. COPA was the only organi-sation which had access to such resources through its national member associations; it did not have the resources itself. In 1975, it only had a staff of 35 (Averyt, 1977, p. 78). The objective method 'initially brought the Commission and COPA together to make the system operate effectively' (Phillips, 1990, p. 55). But after two years of operation, the Commission began to lose faith in the method as the so-called objec-tive calculations turned out to be highly subjective. That is, the results simply de-pended upon who made the calculations (ibid., pp. 55-6). COPA's influence de-clined along with the faith in the method.

Most observers agree that COPA's influence peaked by the mid-1970s (e.g. Grant, 1997; Neville-Rolfe, 1984, p. 247; Phillips, 1990, p. 55; Nedergaard et al., 1993, pp. 50-1). However, COPA remained influential even after that period. Basing his argument upon a close examination of unpublished Commission and Council working group papers, Gardner (1987, p. 170) points out that in the period from 1974 to 1984 'Commission officials ... took a great deal of note of what they were told by COPA representatives.' These papers confirmed the observation of a senior Commission official who said: 'the Commission only proposes and the Agriculture Council will only agree what it knows COPA will accept' (ibid., p. 171).

In the 1980s, the agricultural policy network became looser. One reason for this was that consensus on the Common Agricultural Policy within the Commission broke down. Disagreements among the commissioners became more frequent (Phillips, 1990, p. 53). As Phillips (1990, p. 52) suggests: 'Although deliberations within DG-VI and the Commission are supposed to be confidential, the fact that disagreements have become public suggests the level of agreement has declined.' At the same time, surplus production and budgetary problems of the CAP made life more difficult for COPA. Before these problems came to dominate the agricultural agenda, it could always reach a compromise by increasing price demands, but such a strategy was inadequate in the 1980s. Agricultural policy making was no longer primarily a matter of increasing farmers' income, but also became a matter of de-creasing overproduction and curbing spending increases. As a result, the association had difficulties reaching internal agreement (Tracy 1993, p. 249; Nugent, 1994, p. 367; Neville-Rolfe, 1984, p. 247; Phillips, 1990, p. 73; interview, Danish Council of Agriculture, 1992). In spite of the new agricultural agenda of the 1980s, '[f]or many

[Commission] officials, COPA's arguments have remained unchanged and do not show how they are dealing reasonably with the reality of oversupply, budgetary crises and the demand for lower prices for agricultural raw materials' (Gray, 1989, p. 223, quoted in Grant, 1997). So, COPA is now more of a forum for coordination and for exchange of views and information (Averyt, 1977, p. 72, Gibbons, 1990, p. 379; interview, Danish Council of Agriculture, 1992). However, in situations in which COPA reaches a compromise which DG VI perceives as realistic, it does make an impression on EC officials (interview, DFU, 1992a). Compared to other European interest groups involved in EC agricultural policy making, 'COPA is by far the most influential' (Fearne, 1991a, p. 106; see also Phillips, 1990, p. 83).

COPA's declining ability to present realistic policy proposals resulted in a change in the Commission's strategy. It turned to the Special Committee of Agriculture (SCA) for support for its proposals (Phillips, 1990, p. 52). The Special Committee of Agriculture is an equivalent of the COREPER (the Committee of Permanent Representatives) which is 'a Council' below the ministerial level. By drawing the SCA into the early phases of the policy process, DG VI indirectly strengthened the position of the national farm organisations. Although DG VI has improved its direct contacts with the national agricultural associations (Averyt, 1977, p. 57; Gibbons, 1990, p. 378; interview, Danish Council of Agriculture, 1992), the most important influence is that which these associations achieve through the national civil servants who are members of the SCA. For instance, Danish agricultural associations regard the Danish Ministry of Agriculture and Danish politicians as the most important channels of influence (*Berlingske Tidende*, 31 October 1988; *Børsen*, 20 September 1985). These civil servants are in close contact with their national agriculture ministries. In all EC member states, 'agricultural decisions are taken in a distinct network; highly insulated from the political process as a whole, and usually involving agricultural ministries, farmers' organizations ... [and] specialist legislative committees' (Grant, 1993a, p. 37; see also Keeler, 1996, pp. 138-9). Farmers were not confronted with powerful countervailing interest groups in the agricultural policy networks in any of the member states until the Swedish entry into the European Union in 1995 (Smith, 1990a; 1992 [the UK]; Hendriks, 1991 [the FRG]; Glasbergen, 1992, p. 43 [the Netherlands]; Keeler, 1987; Phillips, 1990, pp. 103-6 [France]; Gibbons, 1990 [Ireland]; Gibbons, 1995 [Spain]; Keeler, 1987, pp. 274-8 [Italy]; Denmark, see the section above). The above analysis shows that, the EC agricultural policy network came to involve more agricultural interest groups in the 1980s. COPA's declining ability to maintain its central position in the agricultural network was used by the Agricultural Directorate to place itself at the centre of the agricultural policy process (Phillips, 1990, p. 71).

Because the agricultural associations have information which the Commission needs effectively to operate the Common Agricultural Policy, farm interests dominate EC agricultural policy making. The Commission needs information on the situation in both domestic and international markets (*Ordbruk* no. 7, August 1994). Assessing demand incorrectly may mean that the Commission has to support expensive market intervention (Collins, 1990, p. 248). Particularly, the management of the policy requires detailed market information on 'harvest levels, qualities, transport difficulties, major projected purchases and the like' (ibid.). Further, In addition, the Commission values estimations on future increases in productivity (Gardner, 1987, p. 170) and national statistics (Nello, 1989, p. 103). The agricultural organisations possess all these types of data and, therefore, they can easily gain access to the policy process through the national agricultural ministries or by influencing the Commission directly. Since neither consumer nor environmental organisations have resources to provide the input the Commission needs, they have not become strong countervailing powers to farmers within the network (Nugent, 1994, p. 366; Collins, 1990, pp. 250-2; Smith, 1990a, pp. 162-8).

In 1973, consumers formed the European Bureau of Consumers' Associations, BEUC, and the year after, environmentalists established the European Environmental Bureau (EEB). Right from the beginning, the EEB focused on agricultural pollution and environmental damage (Phillips, 1990, pp. 73, 92). Neither of the two interest groups have been able to break the dominance of farmers. The application of the objective method in the 1970s implied that farm product prices 'were set totally without regard to consumer [and] environmental ... interests' (Smith, 1990a, p. 155). Thus, from the outset of their appearance in the European Community, both consumer and environmental interest groups belonged to the periphery of the agricultural policy network. Since the Agricultural Directorate to large extent controlled the agricultural policy process within the Commission, BEUC and EEB could not use their close contacts with the Directorate for the Environment, Consumer Protection and Nuclear Safety (DG XI) (ibid., pp. 90, 92) to penetrate the agricultural policy network.

Consumers' attitudes towards the organisation of the agricultural policy process clearly show that they are not core members of the network. They argued in the 1980s that:

> Directorate General VI, responsible for agriculture, behaves as the sponsoring division for farming interests and it fails to take full account of broader Community interests, including those of consumers and taxpayers. It tends to work in isolation from other directorates representing other interests (National Consumer Council, 1989, p. 182).

What is more, consumers claimed that they had no input into the implementation of the CAP, and that by consulting informally, the Commission favoured producer interests (ibid., pp. 182-3). Consumers' critical view on the CAP had little support within DG VI. In the 'green paper' from 1985, the Commission argued that the Common Agricultural Policy had been beneficial to consumers because it had ensured consumers' access to food supplies. The Commission believed that food prices were fair and stable EC countries when compared to those of other industrial countries (Kommissionen, 1985, p. 1; see also Smith, 1990a, p. 166). This point of view was repeated in one of the Commission's reform papers in 1991 (Kommissionen, 1991, p. 1). Moreover, a DG VI official told Smith (1990a, p. 166) that consumer groups had nothing to offer and that they were not representative of their constituencies. A survey conducted in 1987 showed that only 22 per cent of consumers thought that the Common Agricultural Policy was too expensive (Phillips, 1990, p. 90). This information might have strengthened the DG VI's view that consumer groups were not representative of their constituencies. Throughout the history of the CAP, consumers have had only marginal influence. As Ritson (1991, p. 119) points out: 'It is generally accepted that the consumer voice in Europe, in so far as agriculture is concerned, is weak. It is difficult to cite one major example of a CAP decision which has been influenced predominantly by the consumer interest.'

The European food processing industry which is not owned by farmers' cooperatives has not, despite its economic importance and significant interests in the CAP, been able to place itself in a central position within the agricultural policy network (Grant, 1997). CIAA, which represents European food and drink industries, has only gradually become involved in the annual price negotiations. Traditionally, it has been most concerned with food legislation and quality issues (Nello, 1989, p. 103). CIAA has not been an attractive negotiating partner for the DG VI because it has always had difficulties arriving at a common policy position (ibid., p. 104) which the DG VI considered realistic. For example, the confederation has been divided on the question of liberalisation of the agricultural policy. It 'has always been torn between the desire of secondary processors (the producers of what are normally understood as processed or value added foods) to see more free trade in raw materials and the support of primary processors for the CAP' (*AE* no. 1445, 21 June 1991, p. E/3). The basis for resource exchange between the DG VI and the confederation is, however, weak. As Moyer and Josling (1990, p. 48) point out: 'the Commission does not need agri-business support in the same way as it needs farm group support to legitimize its policies.'

Political scientists do not pay much attention to the roles of the budget commissioner, the Budget Directorate and national finance ministers in the agricultural policy process. These play only a very limited role (see e.g. Nedergaard, 1994, p. 100; 1995, p. 139) and thus belong to the periphery of the agricultural policy network. There are two reasons for this. Firstly, as Moyer and Josling (1990, p. 26) point out: 'EC institutions are not well adapted to keeping agricultural spending under control. Integration has not proceeded far enough to create a single decision-making authority which can balance the funding of different community policies.' Secondly, the Council of Finance Ministers does not have the agricultural expertise to counterbalance the influence of the Agricultural Directorate and the Council of Farm Ministers (ibid., p. 35). However, the budget commissioner did play an important role in the introduction of milk quotas and budget stabilisers in the 1980s (ibid., ch. 4). But, in general, there is no evidence that her/his influence is significant. Likewise, the Environment Directorate 'remains a peripheral actor in the formation of the agricultural policy' (Grant, 1997).

The Commission has employed both formal and informal procedures to integrate organised interests into the agricultural policy process. Formal integration is mostly applied in the implementation phase, whereas informal integration characterises the policy formulation phase. There are, however, a few examples of formal integration in the policy formulation phase: in the 1970s, the agriculture commissioner and COPA's presidium met monthly (Averyt 1977, p. 55; Burkhardt-Reich and Schumann, 1983, p. 350). The declining influence of COPA meant that these meetings only took place every fourth or fifth months in the late 1980s (Nello, 1989, p. 103). Another example of formal integration in the policy formulation process is the obligation on the Agriculture Council president to have meetings with COPA and BEUC representatives. While the Council accorded this right to COPA in the early 1970s, BEUC had to wait until the 1980s to achieve the same right (Fearne, 1991a, p. 110). Even though these meetings are mainly symbolic, the fact that consumers achieved status equal with farmers a decade later clearly shows their limited influence.

Contacts between DG VI and COPA are mostly informal. They involve ad-hoc meetings and they are often carried out in telephone conversations (Burkhardt-Reich and Schumann, 1983, p. 350; Nugent, 1994, p. 381). COPA has described its informal contacts with the Commission in the following way:

> As for the more technical issues, regular contacts take place between COPA experts and those of the Commission. These can take various forms - personal contacts at staff level, attendance of Commission officials at COPA/COGECA meetings, transmission of letters and of written positions, and the statements by COPA/COGECA delegates at Advisory Committee meetings (COPA, 1986, p. 8; see also Nedergaard et al., 1993, p. 52)[16]

Until the 1980s, the Agricultural Directorate and COPA consulted each other on a day-to-day basis (Smith, 1990a, pp. 161-2).

As mentioned above, formal integration is more developed in the implementation phase. In 1992, there were 25 advisory committees to advise the Commission on the management of agricultural markets (*Official Journal* L 26, 03/02/92, 92/41/ EEC, Euratom, ECSC: Final adoption of the general budget of the European Communities, section III, part A, appendix A). They are composed of representatives from producer associations, agricultural cooperatives, commercial food processors, traders, workers and consumers. Officials from national associations represent producers and cooperatives. They occupy 50 per cent of the seats, while traders, commercial food processors have 25 per cent, and workers and consumers share the last 25 per cent (Economic and Social Committee, 1980). The main tasks of the committees are to consider the market situation, the Commission's proposal on market management, certain aspects of marketing and technical problems (ibid., p. 9). The advisory committees have no formal powers and, thus, they are of little importance for national organisations (Gibbons, 1990; 383; interview, DDB, 1992). During the 1980s, the number of meetings was cut back, indicating that the importance of the committees had declined. For instance, in the early 1980s, there were 11 cereals meetings and five dairy meetings (Economic and Social Committee, 1980). In 1992, there were only six meetings concerning cereals while the advisory committee for dairy products met only three times (interviews, DDB, 1992 and DFU, 1992a). Consumers do not view the advisory committees as important. They argue that the committees 'appear to offer little more than token consultation - they are poorly serviced and are dominated by representatives of producer interests' (National Consumer Council, 1989, p. 182). Consequently, the Danish Consumer Council no longer attends meetings since the costs of participating far exceed the benefits (interview, Danish Consumer Council, 1993).

The degree of institutionalisation can be established by analysing policy objectives and their long term stability. Article 39 of the Treaty of Rome lays down the policy objectives of the Common Agricultural Policy. They are: to increase agricultural productivity; to ensure a fair standard of living for the agricultural community; to stabilise markets; to assure the availability of supplies and to ensure that supplies reach consumers at reasonable prices (De europæiske Fællesskaber 1987, p. 253-4). Although formally the objectives enjoy equal status, there is no doubt that ensuring that farmers achieve a fair living standard has had the highest priority over the last three and a half decades (Smith, 1990a, pp. 152-3, 206). The Commission has often emphasised the need to protect their incomes. For instance, in 1981, it pointed out that the Community had a large responsibility when it makes agricultural decisions because such decisions had 'an immediate impact on the income situation of eight million employed in agriculture who, with their families, represent forty million people' (Kommissionen, 1981, p. 4, my translation). The 'green paper' repeatedly returned to the question of farmers' income (Kommissionen, 1985, pp. III, 12, 49, 59, 60), arguing, for example, that '[t]he Community must ensure that the social and economic conditions for those working in agriculture do not deteriorate ...' (ibid., p. VI, my translation). Similar concerns were expressed in one of the Commission's reform papers in 1991 which emphasised that price cuts could not be carried out unless the Community compensated farmers for their income losses. Further, it stressed that farmers' income should be safeguarded if the CAP was reformed (Kommissionen, 1991, pp. 8, 12). agriculture commissioner Mac Sharry defended these positions, promising farmers that 'he would continue to protect the idea of compensation for price cuts' (*AE* no. 1490, 8 May 1992, p. E/5).

More or less officially, a major objective of the Common Agricultural Policy is the commitment of the Community to maintain the family farm as the foundation of European agriculture. This goal was written into the Stresa declaration in 1958, stating that 'given the importance of the family farm structure of European agriculture and the unanimous wish to safeguard this character, every effort should be made to raise the economic and competitive capacity of such enterprises' (Commission, 1958, quoted and translated in Hill, 1993, pp. 359-60).

In the 'green paper', the Commission confirmed that 'the family farm is still regarded as fundamental' (Kommissionen, 1985, p. II, my translation) and that 'every effort should be applied to increase the economic capacity and competitiveness of the family farms' (ibid., p. 9, my translation).

The core members of the agricultural policy network, i.e. farmers and DG VI, have not questioned these policy objectives, but have defended them. Consumers, who belong to the periphery of the network, have tried to persuade the Commission to emphasise the clause of the Treaty of Rome's article 39 which states that food supplies should reach consumers at reasonable prices (Fearne, 1991a, p. 106). So far, they have not been very successful in achieving this aim. Indeed, the EC consumer association, BEUC, does not believe that the Community should withdraw completely from the agricultural sector. To pursue social and environmental objectives, the Community should pay farmers direct income support which should be totally decoupled from production (*AE*, no. 1417, 30 November 1990, p. E/3; no. 1442, 31 May 1991, p. E/9; National Consumer Council, 1995, pp. 5, 7-8).[17] Thus it appears that consumers' opposition to agricultural support depends not only on the way it is paid but also on its size.

The high degree of stability in the CAP's policy objectives indicates that the agricultural policy network has a relatively high degree of institutionalisation. Safeguarding farmers' incomes and increasing the economic and the competitive capacity of the European family farms demanded that the EC played a significant role in the agricultural sector. The core members of the network share the view that the Community should intervene in agriculture to protect farmers from the impact of free market forces. In other words, there is a consensus on the principle of state responsibility. In conclusion, the agricultural policy network of the European Community shares certain features of the ideal type policy community. It has a high degree of institutionalisation, a relatively high degree of integration and the members represent a limited range of interests. There is no countervailing power within the network's core. But unlike the ideal type policy community, the network has many agricultural members. They might disagree on the detailed content of policy, but accept the basic policy principles.

Conclusion

In the introduction to this chapter, it was argued that network structures influence the policy preferences of the network's members, in particular those of agricultural authorities. Formation of policy preferences has a crucial influence upon coalition-building in the policy process which, in turn, affects the policy outcome. The theoretical network model suggests that differences in agricultural policy networks help to explain why Swedish farmers were subject to more radical policy changes than farmers in the European Community and Denmark. It has been pointed out that this proposition is valid if an empirical analysis of the agricultural policy networks in Denmark, Sweden and the EC shows that the Danish and the EC networks are more cohesive than the Swedish network. If this pattern is not found, then the model cannot predict policy choices.

The analysis demonstrated that both Danish and EC agricultural policy networks were more cohesive than the Swedish network. While consumers were a strong countervailing power in the core of the Swedish network, the networks in Denmark and the EC excluded countervailing forces from the core. What is more, in Denmark and the European Community there was a consensus on the principle of state responsibility within the core of the network, indicating that the networks had a high degree of institutionalisation. Such a consensus was non-existent within the Swedish network, meaning that the degree of institutionalisation was low. With regard to integration, there were no major differences among the three networks. To conclude, my empirical analysis strengthens the theoretical argument that network structures have important influence on how agricultural authorities position themselves in policy processes in which a political actor not traditionally participating in agricultural policy making demands change.

Notes

1. In using the stability of agricultural policy objectives as an indicator of cohesion, there is a risk of running into problems of construct validity (see the introductory chapter) because change in policy objectives may also be associated with cohesive networks. All members of a cohesive network can agree to alter the policy objectives. However, the problems of construct validity can be overcome because cohesion is also associated with the absence of countervailing forces and a high degree of integration. So, to obtain construct validity, the findings of the analysis of the network's three dimensions should agree with each other. If there is no agreement, there may be problems of construct validity.

2. This section applies data from an interview with an official in the Danish Consumer Council in 1993. It was conducted by Morten Bundgaard Andersen.

3. However, it was not until three years after the establishment of the boards that the ministry became a member (Just, 1992, p. 517).

4. For competitive reasons the Ministry of Agriculture carried out some administration in the butter export board.

5. There is no reason to believe that they were excluded from participation as both the consumers and the labour movement had representatives in the committees.

6. There are now some indications that consumer interests are given higher priority.

7. See also Gad (1969) for a critique of the Danish agricultural policy.

8. In 1934, the cereals regime was changed so that mills were only allowed to use Swedish cereals (Jordbrugsdepartement, 1989, p. 63).

9. Butter exports accounted for 25 per cent of the dairy production (Hellström, 1976, p. 334). The price of butter on the export markets decreased by 40 per

cent from 1928 to 1933 (Seyler, 1983, p. 170).

10. It remains unclear how seriously farmers were hit by the crisis in the 1930s. See Thullberg (1977, p. 197) and Seyler (1983, p. 171) for a discussion on this matter.

11. From 1979 until 1986, the average proportion of the agricultural production value (the PSE percentage) which could be put down to direct and indirect agricultural state subsidies was 70 per cent in Norway. In Sweden, the proportion was 44 per cent and 37 per cent in the EC. In 1990, the proportion had increased to 75 per cent in Norway, 57 per cent in Sweden and 46 per cent in the EC (OECD, 1994, pp. 107-8).

12. See the next chapter for a full account of the alterations in party loyalties.

13. Before 1983, *Sveriges Statskalender* does not list the number of employees in each division.

14. Petersson (1989, p. 61) uses a measure which combines the frequency and intensity of the contacts.

15. These include trade unions and the Cooperative Union and Wholesale Society.

16. The Advisory Committees are set up to advise the Commission on implementation of CAP decisions.

17. According to Peter Nedergaard of the Danish Consumer Council, the positions put forward in this paper represent the positions of BEUC (personal conversation 1995).

8 Farmers' structural power in parliament and state structures

Introduction

In chapter 3, it was argued that two macro-political variables in particular have significant influence upon meso-level policy processes. One is a social group's structural power in parliament and the other is the state structure. The macro- level model suggests that in agri-environmental policy making, the likelihood that environmental interests will succeed in having high cost policies introduced is greatest if farmers have only limited structural power in parliament and if the state is centralised is centralised. High cost agri-environmental policies are based upon the polluter pays principle. By contrast, policy makers are most likely to adopt low cost agri-environmental policies when farmers have structural power in parliament and when the state is fragmented. The transfer of the basic principle of agricultural policy - the principle of state responsibility - to agri-environmental policy making characterises low cost policies. In agricultural policy reform processes, the macro-level model suggests that third order policy reforms are most likely to occur when farmers have limited structural power in parliament and the state is centralised. First and second order policy reforms are most likely to be adopted in fragmented states in which farmers have structural power in parliament.

The empirical analysis in chapter 5 showed that the Danish nitrate policy is a low cost policy whereas the Swedish is a high cost policy. Chapter 6 undertook an analysis of agricultural policy reforms in the European Community (EC) and in Sweden, concluding that the EC adopted a second order policy reform in the arable sector and a first order reform in the animal sector and that the Swedes enacted a third order reform. This chapter examines whether the macro-level model can contribute to an explanation of these differences in policy choices. Since the European Parliament does not have much influence in agricultural policy making, the comparison of farmers' structural power in parliament only includes Denmark and Sweden (see chapter 3). As argued in chapter 3, the EC and states can be compared on the centralisation-fragmentation continuum because the way in which political institutions relate to each other has a major impact on policy change, irrespective of whether the process of change takes place in the EC or in nation states.

Danish and Swedish farmers' structural power in parliament

Farmers' structural power in parliament refers to the extent to which political parties consider agricultural interests in their general policy positions and electoral appeal. Fundamental changes in a policy sector can occur when the structural parliamentary power of the social group which benefits economically and politically from the sectoral policy concerned has eroded. Agricultural interests were well entrenched in parliaments during the formative phases of agricultural policy making in the 1930s. Therefore, only a historical analysis can establish whether the structural parliamentary power of farmers has declined.

The three Scandinavian countries all have multi-party systems with one dominant party. In terms of votes, social democratic (or labour) parties have been dominant. All the countries have at least four other significant parties (Blondel, 1968, pp. 186-9). Although the electoral strength of the social democratic parties has declined, they are still the largest parties in Denmark and Sweden. To establish farmers' structural parliamentary power in a multi-party system, we need to examine the extent to which parties in parliament support agricultural interests. Furthermore, we should examine the parliamentary position of the supporting parties. The analysis involves four steps: firstly, parties which are loyal to farmers are identified; secondly, an assessment of such parties' general loyalty to farmers are made; thirdly, the strategic parliamentary position of such parties is analysed; and finally, the strength of farming interests within the Social Democratic Party is assessed. The analysis concentrates on the critical periods in Danish and Swedish history of agricultural politics.

Denmark

Since the first decade of this century, two parties have represented agricultural interests.[1] In the early 1930s when the agricultural policy network was established, the Liberal Party (*Venstre*) and the Radical Liberal Party (*Det radikale Venstre*) represented farmers' interests. The former, as its English name indicates, holds in liberal ideas, and the latter is a social liberal party. In the 1932 general election, most farmers (about 70 per cent) voted for the Liberal Party, while the Radical Liberal Party got about 20 per cent of the farming votes (Just, 1992, pp. 45-6). Both parties had close ties with agricultural organisations. The Liberal Party was closely connected with the Farmers' Union and the Radical Liberal Party with the Smallholders' Union. Agricultural dominance within the two parties is indicated by the fact that in 1932, two thirds of the Liberal Party's and half of the Radical Liberal Party's members of parliament were farmers or smallholders (ibid., p. 48).

The Liberal Party is still loyal to agricultural interests. It receives about two-thirds of farmers' votes and has close contacts with farmers' associations. In the early 1980s, 71 per cent of the liberal members of parliament were in contact with farmers' associations at least once a month. This was well above average level (25 per cent) of contacts between political parties and interest groups (Damgaard, 1982a, pp. 347, 349). Close links to farmers are also indicated by the fact that in 1994, 21 per cent (9 of 42) of the liberal members of parliament were farmers (*Jyllands-Posten*, 24 September 1994). During the late 1950s and early 1960s, the Radical Liberal Party changed in that it moved closer to middle class voters at the expense of smallholders' interests. Because the smallholders were becoming less important in the national economy, the radical liberals were no longer willing to support them with the subsidies which they required to stay in business (see chapter 7 on the economic problems of agriculture in the late 1950s). In this situation, the Radical Liberal Party chose to emphasise national economic efficiency over the interests of the smallholders (Søborg, 1983, pp. 213-14, 237-8). In 1961, the radical liberal Minister of Economic Affairs argued that it would be better for the state to subsidise farmers rather than smallholders (Larsen ed., 1980, p. 47)[2] There was another indication that the Radical Liberal Party was detaching itself from the smallholders. Traditionally, the leader of the party organisation was always a farmer or a smallholder, but that tradition was broken in 1964 (ibid., p. 55). The decrease in the number of radical liberal members of parliament who were farmers also weakened agricultural interests within the party. In 1966, there was only one (Pedersen, 1968, p. 7) and since 1990 there has been none. Despite the detachment from smallholders' interests, the radicals were still able to attract farmer and smallholder votes. During the 1960s, the parties' share of the farming votes even increased: it was 8 per cent in 1959, 22 per cent in 1968, and 20 per cent in 1971 (Larsen ed., 1980, p. 136).

In 1982, a liberal conservative four-party government consisting of the Liberal Party, the Conservatives, the Centre Democrats, and the Christian People's Party was formed. Compared to earlier governments, this one was unique because it had to face an alternative parliamentary majority in some policy areas, particularly in security policy making but also in legal affairs, environmental, energy and cultural policy making.[3] The radical liberals belonged to that majority (Damgaard, 1992, p. 34). Being part of the alternative majority in a period when pollution control in agriculture was on the agenda demonstrated that the party emphasised environmental interests over farmers' interests. However, in agri-environmental policy making this position caused an internal conflict between a member of parliament representing smallholders' interests and the urban middle class wing of the party (Andersen and Hansen, 1991, p. 71).

Parties supporting farmers have held strategic parliamentary positions during critical periods in Danish agricultural policy making. In 1929, a majority coalition government was formed by the social democrats and the radical liberals. It held office until 1940 (Damgaard, 1992, pp. 24-5). Since the social democrats depended upon the support of the Radical Liberal Party, the latter was in a strategic position. Until 1936, the government commanded a majority only in the lower chamber. Thus, to pass legislation in the upper chamber,[4] it depended on support from the Liberals or the Conservatives. Hence, the Liberals were in a strategic position in the formative phases of Danish agricultural policy making in the early 1930. The Liberal Party reached a compromise with the government on the content of agricultural policy and on the design of the agricultural policy network (Just, 1992, ch. 2-4).

In the decades to follow, the Radical Liberal Party maintained its central role in Danish politics. It had considerable influence on the creation and the fate of social democratic and liberal conservative governments. As Pedersen (1987, pp. 9-10) points out: 'The Radical Liberals mostly found themselves placed in a pivotal situation after the Folketing elections so they were bound to become mediators and power-brokers in Danish politics.' During the period 1982 until 1988 in which the *Folketing* enacted two agri-environmental action plans, the Radical Liberal Party (which was not one of the government parties) 'was pivotal in most cases of conflict' (Damgaard, 1994, p. 93). The Liberal Party has not had a similar strategic position; however, during the periods of liberal conservative governments in the 1980s, the party was able to ensure that agricultural interests were not disregarded by the other government parties.

In terms of votes, the Social Democratic party has been the dominant party in Denmark since 1929. It has had an important influence on agricultural policy. From the very beginning of the party's existence, its leaders realised that it could not come to power if it based its electoral support on urban workers alone. It had to win the support of farm workers and parts of the rural middle class, that is farmers and smallholders. Consequently, the social democrats chose a moderate line in its 1882 agricultural policy programme. They even regarded farmers as a part of 'the working people.' Farmers were definitely not conceived of as capitalists, although most of them hired farm workers. The social democrats based this view of farmers on the fact that farmers themselves, through manual labour, took part in the production process (Bryld, 1992, pp. 240-4). The moderate agricultural policy programme did not conflict with the interests of the workers who favoured low food prices. Conflicts over food prices might have emerged if Danish farmers had not responded to the imports of cheap American cereals in the 1870s by transforming from vegetable to animal production. By so doing, Danish agriculture managed to remain competitive in the world market. Since the conflict over food prices was absent, the Social Democratic party did not become a strong voice of consumer interests.

State intervention in agriculture in the 1930s was not controversial for the social democrats even though consumer and labour interests were not included in the agricultural policy network which was established. According to Just (1992, p. 43), agricultural policy making was apparently a non-issue in the party and was left to the social democratic minister of agriculture to decide. Agricultural interests were relatively strongly represented in the party as one-eighth of its members of parliament were farmers (ibid., p. 48). The party only attracted about 2 per cent of farmers' votes (Thomsen, 1987). Although the agricultural policy raised domestic prices, there was only weak opposition to the policy within the Social Democratic Party. One of its leading members of parliament was critical of the policy but he was not able to prevent it from being put into place (Hansen and Torpe, 1977, p. 178).

During the 1980s and early 1990s, the Social Democratic Party's sympathy for

175

farmers' interests declined a little. In particular, agri-environmental and salmonella problems moved the party closer to environmental and consumer interests. For example, one of its then leading members of parliament, Ritt Bjerregaard,[5] attacked farmers for being 'environmental pigs' (Madsen, 1987, p. 105, my translation). Later, she strongly criticised farmers for doing too little about the salmonella problem (*Aarhus Stiftstidende*, 20 June 1993). One can easily over-emphasise the decline in the Social Democratic Party's support for agricultural interests in the 1980s and early 1990s. For the party, the membership of the alternative parliamentary majority in the 1980s was motivated more by the opportunity to force the liberal conservative governments to resign (Andersen et al., 1992, p. 9) than by waning sympathy for farmers. Since the social democrats returned to power in 1993, the ministers of agriculture and the environment have been careful not to enter into severe conflicts with farmers, indicating that the purpose of supporting environmental and consumer interests in the 1980s and early 1990s was to overthrow the liberal conservative governments. In relation to agricultural policy, the social democrats have traditionally been concerned about employment in the food processing industry. National economic interests and agricultural interests have always been closely related in Denmark (see chapter 7). But as industrial production increased, the relative economic importance of agricultural production declined. This did not significantly change the social democrats' responsiveness to agricultural interests. That employment and national economic concerns associated with agricultural production are important for the party became evident when its agricultural spokesman said in 1996 that he did not want his party to be seen as being negative towards agriculture. He continued: 'I know what agriculture means for export incomes and the employment' (*Landsbladet* 15 March 1996, my translation). Attempts to get the farmer's votes has not been the main driving force behind the Social Democratic Party's agricultural policy positions in recent time; only 1 per cent of farmers voted for the Social Democratic Party in the 1988 general election (Glans, 1989, p. 68).

To conclude, over the last six-and-a-half decades, farmers' structural power in parliament has declined a little. The Radical Liberal Party has emphasised urban middle class interests more than those of smallholders and the Social Democratic Party has been slightly more responsive to environmental and consumer interests.

Sweden

Traditionally, Swedish farmers have been represented in parliament by the Agrarians (*Bondeförbundet*) which later became the Centre Party (*Centerpartiet*). In the 1930s, the party had a very marked agrarian profile but it also appealed to the rest of the rural population. Farmers' interests were seen as being equal to national interests (Larsson, 1980, pp. 143-5). In parliament, farmers' interests were well represented; between 89-97 per cent of the agrarian members of parliament were farmers (Micheletti, 1990, pp. 32-3).

From the mid-1940s, the party began to disassociate itself from farmers' interests. The 1946-programme put more emphasis on general rural interests than earlier programmes had done. During the 1950s, the party tried to appeal to new groups of voters but it was not until the late 1950s, that it took a symbolic step away from farmer interests. In 1957, its name was changed to the Agrarian Centre Party (*Centerpartiet Bondeforbundet*). One year later, its official name became the Centre Party. The defeat in the 1956 general election and the steady depopulation of the countryside motivated a group of primarily urban and well-educated members to take the lead in the transformation from a rural and agrarian party to a party appealing to both urban and rural voters (Larsson, 1980, pp. 115-46; Steen, 1985, pp. 58-9, 1988, pp. 339-40). The change in party profile is clearly indicated by the contents of the party programme. From 1956 to 1960, the share of the programme which dealt with agriculture fell from 37 to 6 per cent. The emphasis on environmental protection in the mid-1970s also shows that the party was appealing to new social groups (Christensen, 1994, p. 332-3). Farmers' influence declined in the Centre Party's parliamentary group: in 1965, 57 per cent were farmers and in 1985, the percentage had declined to 28 per cent (Steen, 1988, p. 293; Holmberg and Esaiasson, 1988, pp. 141-2). The proportion of Centre Party members of parliament who regard themselves as representing agricultural interests has also decreased. In 1969, 21 per cent of the party's members of parliament said that they represented agricultural interests whereas in 1985, only 15 per cent believed that farmers' interests were most important (Holmberg and Esaiasson, 1988, pp. 36, 40).

Because it is a potential coalition partner for the other parties in the *Riksdag* (the Swedish parliament), the Agrarian/Centre Party has always had a central position in Swedish politics. In 1933, the social democrats made a historical compromise (*krisuppgörelsen*) with the Agrarian Party and in 1936, the two parties formed a coalition government which was in power until the outbreak of the Second World War.[6] The Agrarian Party entered a coalition government with the social democrats in 1951 which lasted until 1957. After the breakdown of the coalition, the Centre Party successfully placed itself in the centre of Swedish politics as an alternative to both the social democrats and the two non-socialist parties (the Liberal Party and the Conservative Party), refusing to join any of the two blocs. By being in the centre of Swedish politics, the Centre Party has benefited from other parties' attempts to co-operate with it. Since the social democrats lost the unconditional support of the communists in the 1980s, the Centre Party has strengthened its strategic position (Sannerstedt and Sjöblom, 1992, pp. 102-7; Micheletti, 1990, pp. 100-1).

The Social Democratic Party formed its position in agricultural policy making very early. In the late 1870s, cheap cereals imports from Russia and the United States forced the domestic prices down and triggered a major agricultural crisis. As other European countries, Sweden had to choose between a protectionist or a free trade policy. This question awakened the political interest of the Swedes, demonstrated by the increase in electoral participation: it almost doubled in the 1887 general election (Carlsson, 1979, pp. 95-6). Put simply, producers of cereals pursued a protectionist line, whereas urban workers favoured free trade.[7] Tariff policies were eagerly discussed in workers' clubs as well as in other political associations. Eventually, the protectionists gained a majority in the *Riksdag* which they used to introduce tariffs on agricultural and industrial products (Lewin, 1984, ch. 2; Carlsson, 1979).

Having its electoral base in the working class, the Swedish Social Democratic Party chose a free trade position in agricultural policy making. However, in the 1920s, the party gradually became more in favour of protectionism in order to gain electoral support from new groups, primarily smallholders and farm workers. To obtain a parliamentary majority for their social reform programme, the social democrats depended on the votes of these groups. Although the party tried to balance consumer and agricultural interests, the latter came to dominate agricultural policy making (Thullberg, 1974, pp. 130-47). Since members of parliament who were themselves farmers dominated agricultural policy making within the party (ibid.), the prominence of agricultural interests was not surprising. These members of parliament held between 12 and 15 per cent of the social democratic seats in the *Riksdag's* lower chamber from 1922 to 1937. Concern for farmers' interests was also furthered by the social background of the other social democratic members of parliament. More than a third of those in the lower chamber lived in the countryside (Micheletti, 1990, pp. 32, 36). To cope with the agricultural crisis of the 1930s, the social democrats gave up what was left of their free trade position in agricultural policy. In the historical compromise with the Agrarian Party in 1933, they traded off consumer interests for the agrarians' support for a new labour market policy (Thullberg, 1974, pp. 162-4).

It was not until the late 1950s and early 1960s, that the Social Democratic Party broke with its view of agricultural policy and returned to its former consumer friendly position. In the 1960 programme, the party abandoned support for small-holders and began to emphasise efficiency and the benefits of large scale farming in order to ensure reasonable food prices for consumers (Steen, 1988, p. 310). This change in policy position is associated with the decline in the number of social democratic members of parliament who represent farmers' interests. While there is a good reason to assume that at least 12 to 15 per cent represented farmers in the 1930s, only 2 per cent in 1969 and 3 per cent in 1985 said that they represented agricultural interests (Holmberg and Esaiasson, 1988, pp. 36, 40). The decline in sympathy for agricultural interests is not surprising when one looks at the impor-tance of the farmer vote. In 1956, 4 per cent of the social democratic voters were farmers. The share fell to 1 per cent in 1965 and remained at that level until the late 1980s (Steen, 1985, p. 59; 1988, p. 287, Birgersson and Westerståhl, 1992, p. 57). The decision to include consumers in the agricultural policy network clearly re-flected the new emphasis on consumers interests. That decision was initiated by the social democrats who worried about the imbalance in favour of agricultural interests and therefore wanted to strengthen consumer interests (Steen, 1988, pp. 214-15). The social democrats 'relate[d] price increases and government subsidies to con-sumer interests. [The costs of agricultural policy] were defined as being a question of redistribution between farmers and consumers' (Steen, 1985, p. 52).

In conclusion, Swedish farmers' structural power in parliament has declined considerably since the 1930s. The Centre Party changed from an agrarian party into one which also represented middle class voters. Furthermore, since the late 1950s, the Social Democratic Party has emphasised consumer interests over farmers' interests. From the early 1930 until the late 1950s, it was more concerned with small-holders' interests.

Farmers' structural power in parliament and the choice of policy

The theoretical macro-level model suggests that change in the agricultural sector is associated with alterations in farmers' structural power in parliament. A high degree of structural power makes it unlikely that policy makers will adopt high cost agri-environmental policies and third order agricultural policy reforms. Such policies are more likely to be introduced when farmers have only limited structural power in parliament.

This empirical analysis has showed that Swedish farmers' structural power in the *Riksdag* has declined considerably since the early 1930s. This contributes to explaining why Swedish nitrate policy is a high cost policy and why the 1990 agricultural policy reform was a third order reform. Danish farmers' structural power in the *Folketing* has also eroded but not to the same extent as in Sweden. Therefore, Danish farmers could mobilise sufficient opposition against the Environmental Protection Agency's attempt to introduce a high cost nitrate policy in agriculture. The limited decline in farmers' structural parliamentary power thus partly explains the policy choice in Denmark.

A comparison of state and European Community structures

The other macro-political variable which has explanatory force at the meso-level is the structure of the Danish and Swedish state and the EC. Because actors belonging to centralised political systems have good opportunities to coordinate their actions they can generate authority and, thus, develop into powerful actors. Such actors may have the power to bring about fundamental changes in various policy sectors. Political systems in which authority is dispersed are likely to develop competing decision making centres since coordination is difficult. This makes it difficult for an actor to generate sufficient power to change the established order in a policy sector.

Can differences in state structure explain why Danish and Swedish nitrate policies are fundamentally different? Both states are unitary states. From a comparative perspective, Denmark and Sweden have relatively strong central governments although central power 'is balanced by an extensive system of local government' (Lane and Ersson, 1991, p. 219). Damgaard (1994, p. 86) characterises the Scandinavian system of government as a 'unified system of government' in which, 'the executive is not separately elected ... [and] the executive is responsible to the legislature.' One could argue that the Swedish state is more centralised than the Danish. The former seems to provide the best opportunities for coordination because, in principle, the Swedish government is a collective body, meaning that the ministers, in most cases, cannot make decisions unless the cabinet as a whole approves them.

179

The Swedish principle of governance is unique (Larsson, 1986, p. 283; Lindblad et al., 1984, p. 161; Petersson and Söderlind, 1992, p. 54). In Denmark, the governance principle is 'ministerial governance' (*ministerstyre*) (Lindblad et al., 1984, p. 161). This means that ministers have a high degree of autonomy in relation to the other cabinet members (Christensen, 1980, p. 74; 1985, p. 114; Knudsen, 1993, p. 138). Generally, a minister's relationship to the *Folketing* is more important than his responsibility to the cabinet (Knudsen, 1993, p. 139). Formally, even the prime minister has limited opportunity to intervene in the ministers' businesses.

However, there are factors which makes it unlikely that the Swedish state is more centralised than the Danish. Firstly, despite the principle of ministerial governance in Denmark, coordination does take place among ministers within a network of cabinet committees (Christensen, 1980, pp. 125-97; 1985; Christensen and Ibsen, 1991, pp. 66-74). Secondly, although the Swedish government is formally a collective body, it is not clear whether, in practice, it is better to coordinate than governments of other countries. On the one hand, the Swedish government makes more collective decisions than other governments; 20,000 to 25,000 annually. Most of its ministers make only few hundred decisions independently. In most countries, the reverse is the norm. Furthermore, compared to other countries, Swedish ministers meet with each other more often (Larsson, 1993, pp. 189, 213). On the other hand, most of the 25.000 collective cabinet decisions are just formal; several hundred decisions are often made within half an hour. Hence, Larsson (1993, pp. 200, 282-3) concludes that the Swedish cabinet is at least as segmented as other European cabinets; indeed, the coordination process looks like those of most other European governments (Larsson, 1986, p. 283). Thirdly, in Sweden, the constitution separates politics and administration, meaning that ministers cannot intervene in the administrative agencies' day-to-day administration of laws (Petersson and Söderlind, 1993, p. 74). In contrast, a Danish minister is the political head of administration within her/his ministry's jurisdiction, meaning that s/he can intervene in the agencies' day-to-day administration (Christensen, 1980, pp. 79-81). This difference in the relationship between ministers and administrative agencies shows that, formally, the Danish state is more centralised than the Swedish in this respect. However, in practice, there may not be much difference. Petersson (1989, pp. 67-86) shows that there are close contacts between Swedish government departments and the administrative agencies, but his data does not indicate whether these contacts concern information exchange or are attempts to control the behaviour of the agencies. To sum up, the available research of state structures in Denmark and Sweden does not allow us to argue that Sweden is more centralised than Denmark. Thus, the analysis of Danish and Swedish state structures does not contribute to the explanation of different nitrate policy choices. Indeed, further research and comparisons are necessary before we can draw firm conclusions. Differences in farmers' structural power in parliament is a much more important macro-political variable explaining the variation in policy choices.

Agricultural policy reform in Sweden and in the European Community produced different policy outcomes. Can differences in the EC's and Sweden's governance structures explain this? As a unitary state, Sweden has a relatively high degree of centralisation. How does the EC compare to Sweden in centralisation?

Compared to international organisations and to federal states, the EC is a unique construction (Keohane and Hoffmann, 1991, p. 13). It is neither an international organisation nor a federal state (Brewin, 1987, p. 2; Nørgaard, 1994, pp. 245-6; Peterson, 1995, p. 84). As Keohane and Hoffmann (1991, p. 12) argue: 'the European Community by no means approximates a realistic image of a modern state.' They proceed with saying that when its authority is compared with that of 'contemporary international organizations the Community looks strong [but] in comparison with highly institutionalized modern states it appears quite weak indeed.' Compared to other international organisations, the European Community is strong because it 'as a whole has gained some share of states' sovereignty' (ibid., p. 13). When the EC is contrasted with a unitary state like Sweden, it is evidently that it is highly fragmented. In fact, the EC is no more than a union of states without unity of government (Brewin, 1987). Before the enlargement in 1995, the EC consisted of, at least, 13 decision making centres: the 12 nation states and the Commission. The 12 nation states are strong decision making centres, each capable of adopting national policies to supplement EC policies if they view such a move as being in their national interest. In agricultural policy making, for example, the Commission has feared that policy decisions which did not accommodate the national interests of all member states could lead to renationalisation of the Common Agricultural Policy (CAP) which, in turn, would have serious consequences for European integration (Kommissionen, 1985, pp. IV, 8; Kjeldahl, 1994, p. 21; Grant, 1995). The member states' capabilities for supplementing the CAP with national agricultural policies are clearly reflected in the percentage of net value added in agriculture which can be derived from national support schemes. In 1991, the EC average was 13 per cent. This figure includes great variation between the member states: in Germany it was more than 30 per cent (the highest) whereas in Denmark it was only about 2 per cent (the lowest) (Hansen, 1993, p. 159).

Voting rules in the Council of Ministers are also an important fragmenting factor in the European Community. Before the Single European Act came into force in 1987, few decisions could be taken by majority vote. Many decisions could not be made unless the ministers achieved unanimity. The 'Luxembourg compromise' from 1966 strengthened the rule of unanimity, declaring 'that where 'very important interests' were at stake, the Council would endeavour to reach solutions which could be adopted by all members of the Council' (Tracy, 1989, p. 264; see also Fearne, 1991, p. 32). In practice, the 'Luxembourg compromise' meant that a member state could veto a proposal which conflicted with its vital national interest. The Single European Act extended the use of qualified (weighted) majority voting.[8] This development has raised questions as to whether the 'Luxembourg compromise' is still in force. It has not been totally abandoned. For instance, in 1988, the Greeks successfully threatened to veto the CAP price package (Swinbank, 1989, pp. 312-13), the French government declared in 1992 that it could still be used (Swinbank, 1993, p. 365), and the Italians unsuccessfully tried to block the enactment of the 1992 CAP reform by threatening to apply a veto (AE no. 1492, 22 May 1992, pp. P/2, P/13).

However, in agricultural policy making, the Single European Act implied that 'it has become more difficult in matters of agricultural policy to openly apply a veto. Nevertheless, the threat of blockage remains, and the tendency in the Agricultural Council is still to accommodate a Minister who is opposing an agreement rather than to vote him down' (Tracy, 1989, p. 334).

Although the Commission is a collective body, it divides along national lines. Until the mid-1960s, the Commission consisted of commissioners who had been there since the creation of the EC and had idealistic attitudes towards European integration (Averyt, 1977, pp. 84-6). Thereafter, a new generation of commissioners took over. These '[c]ommissioners and their staffs [were] increasingly conscious of their separate national identities and loyalties' (ibid., p. 84). Compared to the first generation of commissioners, the commissioners of the early 1990s were also more nationalistic. For instance, the Commission was split along lines of national interest in the 1992 agricultural reform process (*AE* no. 1424, 25 January 1991, p. E/3).

So, in the EC, authority is dispersed. Compared to Sweden, the EC is considerably more fragmented. This contributes to explaining why Swedish reformers' ability to bring about a third order reform in Swedish agricultural policy in 1990, and why the 1992 agricultural policy reform in the EC was much more moderate. The Swedish state structure made it possible for the reformers to generate authority and, thus, become powerful actors. The limited opportunities for fundamental policy reforms in the EC meant that the question of third order policy reform was not even raised.

At first sight, the highly fragmented structure of the EC and the presence of an agricultural policy network which shares many features of policy community seem to contradict the macro-level model which suggests that in fragmented or states, issue networks prevail. However, the model is flexible enough to explain why a tight and rather closed agricultural network emerged in the EC. Although the agriculture commissioner and the Agricultural Directorate (DG VI) were operating within a highly fragmented structure when the CAP was formed in the 1960s, they succeeded in becoming hegemonic actors in agricultural policy making (see chapter 7). Thus, they could set up a policy network which excluded the other directorate generals and interest groups which questioned the whole idea of the CAP. The DG VI had good opportunities to achieve a dominant position because it, in the Commission, alone had the agricultural expertise. Another factor facilitating the formation of a relatively tight and closed policy network was that national agricultural policy making rated agricultural interests higher than those of consumers (see chapter 7 and Hendriks, 1991, p. 161). Furthermore, the original six member states did not resist the DG VI and the agriculture commissioner's dominance within the Commission. However, since member states ensured a certain degree of national autonomy by enacting the 'Luxembourg compromise', the EC remained fragmented, making life difficult for reformers.

Conclusion

The theoretical model developed in chapter 3 suggests that particularly two macro-political variables can help to explain policy change in the agricultural sector. Change is associated with alterations in farmers' structural parliamentary power and with state structures (and the structure of the EC). To strengthen the external validity of the model (see the introductory chapter), Swedish farmers' structural power in parliament must be lower than that of Danish farmers. The model cannot otherwise help to explain why the Swedes adopted a high cost and the Danes a low cost nitrate policy. The comparative analysis in this chapter showed that Swedish farmers' structural power in the *Riksdag* has decreased considerably since the early 1930s when the agricultural policy was formed. Danish farmers' structural power in the *Folketing* eroded only slightly. Decline in the Swedish farmers' structural parliamentary power also helps explain why the Swedish 1990 agricultural policy reform was a third order policy reform. Since these finding fit the model's predictions, the external validity of the theoretical model has been strengthened.

The empirical findings, to some extent, strengthen the external validity of the model with regard to state (and EC) structures' influence on policy choices. For the model to be valid, Sweden must be more centralised than the EC and Denmark. There is no doubt that Sweden is more centralised than the EC, but it is not possible to show any significant difference in the degree of centralisation when Denmark and Sweden are compared. Therefore, the model is not detailed enough to reveal variations among countries which can be placed in the same broad categories of political systems, for example unitary or federal states. But when we compare broad categories of political systems, the model does possess explanatory force.

Notes

1. In the 1930s, farmers' interests were also represented by a third party, the Agrarian Party (Bondepartiet). The party gained five seats in the *Folketing* in the 1935 election, but after the Second World War it disappeared from the political scene (Just, 1992, p. 44).

2. This argument caused some annoyance within the party (Larsen ed., 1980, p. 48).

3. From 1982 to 1988, the government accepted 105 defeats in parliament, which is an unusually high number in Denmark. Previously, the norm was that the government resigned or called an election on the issue on which is was defeated (Damgaard, 1992, pp. 31-4).

4. The upper chamber was abolished in 1953.

5. She is now the EU environment commissioner.

6. During the war, Sweden had a grand coalition government consisting of the four major parties.

7. However, workers in the textile industries did not support free trade (Carlsson, 1979, p. 95).

8. Until 1995, a qualified majority required at least 54 votes. The total number of qualified votes was 76.

9 Conclusion: a defence of policy network analysis

A major purpose of this book has been to establish whether policy network analysis can help to explain why agri-environmental policies and agricultural policy reforms differ. The theoretical chapters argued that policy networks have a crucial influence on the choice of policy. To improve the understanding of meso-level policy choices, this book supplemented network analysis with a macro-level analysis of the broader political context. The empirical analysis showed that we cannot neglect the role of policy networks if we seek to understand the choice of nitrate policies in Denmark and Sweden and the outcomes of agricultural policy reforms in the European Community (EC) and Sweden.

During the 1980s and the early 1990, pressure for change arose outside agricultural policy networks. Environmental interests, in particular the national environmental protection agencies, put agri-environmental policies on to the agenda in Denmark and Sweden. Pressure to reform the Common Agricultural Policy came from, particularly, the United States in the GATT Uruguay round. In Sweden, both the minister and the Department of Finance were eager to reform the agricultural policy. None of these outsiders had central positions in the agricultural policy networks. Faced with pressure which was similar in type and intensity, Danish and Swedish policy makers reacted differently. They adopted a low cost nitrate policy in Denmark and a high cost nitrate policy in Sweden. The former is characterised by the transfer of the principle of state responsibility from the agricultural policy to the nitrate policy. It uses policy instruments which pass political and economic costs on to groups other than farmers, and its objectives do not contradict established agricultural policy objectives. The polluter pays principle underpins high cost nitrate policies. This means that policy instruments which concentrate costs on farmers are employed, and the objectives of the nitrate policy contradict agricultural policy objectives.

Agricultural policy reforms in the European Community and Sweden also differed. The EC adopted a second order policy reform in the arable sector and a first order reform in the animal sector. Sweden enacted a third order reform. First order reforms are characterised by alterations in policy instrument settings (for instance adjustments in agricultural minimum prices). Instruments, policy objectives and policy principles remain the same. In second order reforms, instruments and objectives are changed; policy principles are maintained. Third order reforms are much more fundamental: instruments, objectives and policy principles are all changed.

The theoretical models suggest that policy makers have opportunities to bring about fundamental policy change, i.e. third order agricultural policy reforms and/or high cost agri-environmental policies, when a non-cohesive policy network exists in the agricultural sector, when farmers' structural power in parliament has declined, and when the state is centralised. Policy change is likely to be only moderate if these conditions are absent. Thus, when the agricultural policy network is cohesive, when farmers have maintained structural power in parliament, and when the state is fragmented, policy makers tend to adopt first or second order agricultural policy reforms and/or low cost agri-environmental policies.

The comparison of agricultural policy networks supported the theoretical models. One important reason the Danes chose a low cost nitrate policy and the Swedes chose a high cost policy was that the Danish agricultural network was much more cohesive than the Swedish. This meant that the Danish Ministry of Agriculture and farmers could form a strong coalition against environmental interests. Such a coalition could not be formed in Sweden. Farmers therefore became isolated in their attempts to resist environmental interests. The Danish network is the more cohesive mainly for two reasons. Firstly, while the core of the Danish network consisted only of the Ministry of Agriculture and the agricultural associations, the Swedes have integrated the farmers' counterpart in the market, the consumers, into the network. Secondly, there was a consensus on the principle of state responsibility among the Danish network's core members. This consensus was absent in Sweden. In explaining agricultural policy reforms, the comparison also supported the network model. Compared to the Swedish agricultural network, the EC network is the more cohesive. In the latter, there were no centrally positioned countervailing interests. Its core consisted of farmer associations (COPA, COGECA and national agricultural organisations), the agriculture commissioner and the Agricultural Directorate (DG VI). Furthermore, a consensus on the principle of state responsibility had developed.

Sectoral policy choices must be explained in the broader context within which meso-level processes take place. The empirical findings supported the macropolitical model suggesting that state structures and farmers' structural power in parliament have a crucial influence on policy choices. One reason Denmark applies a high cost nitrate policy is that, over time, farmers have succeeded in maintaining strong structural power in parliament. However, this has declined somewhat since the early 1930s. The Liberal Party, which has traditionally represented farmers' interests, is still very loyal to them. Smallholder interests have traditionally been represented by the Radical Liberal Party, but since the late 1950s and early 1960, the party has placed more emphasis on representing urban white collar interests. Even though the Social Democratic Party has moved towards consumer and environmental interests, it is still concerned about employment in the food processing industry and about agricultural export incomes. Therefore, the party is careful not to pursue policy objectives which conflict with basic agricultural interests. Swedish farmers' structural power in parliament has declined significantly since the early 1930s. Traditionally, the Centre Party (the former Agrarian Party) supported agricultural interests. Its transformation from an agrarian party to a party which also represented urban middle class interests weakened farmers' structural power in parliament. Furthermore, the Social Democratic Party which sympathised with agricultural interests from the 1920s until the late 1950s, now promotes consumer interests over those of farmers. The erosion of farmers' structural power in parliament is a major reason the Swedes adopted a high cost nitrate policy. Since both Denmark and Sweden are unitary states with high degrees of centralisation, differences in state structure could not help to explain why Danish and Swedish nitrate policies differ.

The decline of farmers' structural power in parliament also helps to explain why Swedish policy makers adopted a third order agricultural policy reform. Since the European Parliament has no legislative powers and can only express its opinion in agricultural policy making, there was no point in comparing it with the Swedish parliament. The European Parliament is not, therefore, included in the macro-political analysis. When one compares the structure of the European Community with that of the centralised Swedish state, it is apparent that the EC is highly fragmented. Before the enlargement in 1995, the EC in reality consisted of 13 decision making centres - the Commission and the 12 nation states. Even the Commission itself is divided along lines of national interests. Fragmentation results in coordination difficulties with the effect that competing decision making centres are likely to emerge. This, in turn, may produce deadlock. No actor will be able to generate sufficient authority to bring about fundamental policy change.

The theoretical models include the influence of meso-level policy networks, state structures and interest groups' structural power in parliament. It is important to point out that they do not include all the factors influencing opportunities for change. Nevertheless, the models do explain major conditions for change despite the fact that they are relatively simple. The models address the *opportunities* for change; however, even though the conditions for change are present, there need not be change. Reformers and policy innovators must realise that there are favourable opportunities for the pursuit of their interests before they act. They may not be aware of these. As Rhodes and Marsh (1992a, p. 195) point out: 'Actors in the network shape and construct their 'world', choosing whether or not, and how, to respond.' To predict when actors use their opportunities requires a micro-level analysis. We, then, need to consider how individuals who lead and represent organisations conceive of the situations they confront. No doubt, such actors 'are intentional, interpretative beings' (Marsh and Smith, 1996, p. 17; see also Nørgaard, 1996, p. 38). They interpret events but not irrespective of network structures and the broader macro-political context. Thus, individuals who represent organisations which are members of a policy network interpret change arising outside the network through their own understanding and 'in the context of the structures, rules/norms and interpersonal relationship within the network' (Marsh and Smith, 1996, p. 21). Network analysis thus needs a micro-foundation which considers how individuals representing organisations within networks interpret events and how these interpretations influence the actions of their organisations. Such a micro-level model must also consider whether different types of policy networks have different impacts on individuals' perceptions. Finally, a micro-level model must consider also how individuals interpret the actions of those representing other organisations within and outside networks.

Using policy network analysis is not the only way to analyse the relationship between state actors and interest groups in public policy making. One can, for instance, use various institutionalist approaches (see e.g. Hall and Taylor, 1996, Nørgaard, 1996 for a review of these), the public choice approach (e.g. Olson, 1965) or the advocacy coalition framework (Sabatier, 1986; 1993). The next step for policy network analysts is to compare these or other approaches with policy network analysis and subsequently assess the fruitfulness of policy network analysis. Such a comparison may reveal opportunities for combining the approaches.

Do the theoretical models reach beyond the agricultural policy sector? The research design was chosen so that the study would produce the most convincing results with regard to generalisation. The book has applied a comparative case study method in which I based the comparisons of nitrate policy making and agricultural policy reforms on theoretical replication. The logic of that method is that the researcher selects cases, which for reasons predicted by the theoretical models, produce different outcomes. Comparative case studies based on such a research design produce the most robust results (Yin, 1989, pp. 52-54, 109-12). As argued above, the comparative case studies supported the theoretical models. Therefore, there are reasonable grounds to conclude that the models are valid in the agricultural sector.

Although there are good reasons to believe that this study has produced results which are valid in agriculture, it does not follow that the theoretical models are valid in other policy sectors. Only case studies of non-agricultural policy sectors which have been subject to public regulation for a number of years and are facing outside pressure for change can strengthen the general validity of the model. That one cannot *a priori* claim validity outside agriculture does not mean that the models are useless in other policy sectors. Clearly, they are applicable as analytical tools outside agriculture.

This book has shown *that policy networks structures make a difference*; however, sectoral policy choices cannot be explained in isolation from the broader political context. Marsh and Rhodes (1992, p. 2) point out that 'the existence of a policy network both has an influence on, although it does clearly not determine, policy outcomes *and* reflects the relative status, or even power, of the particular interests in a broad policy area.' This study demonstrated that when policy change is put on to the agenda, the type of network which exists in the policy sector subject to change has an important influence on the choice of policy. Network structures strongly influence coalition-building within the network which, in turn, affects the opportunities for outsiders pursuing policy change. Where a cohesive network exists, policy change is likely to be only moderate because network members can form a strong coalition against outsiders. By contrast, radical policy change is more likely to occur where a non-cohesive network exists. In such a network, members have difficulties forming strong coalitions.

The question of who has power within a policy field is also significantly influenced by the network. For instance, a comparison of Danish and Swedish farmers shows that the former are the more powerful, because the Danish agricultural policy network enables farmers to mobilise the support of the Ministry of Agriculture when outsiders threaten the established order in the agricultural policy sector. Indeed, the ministry did not need to be mobilised by anyone in the 1980s when environmental interests tried to introduce agri-environmental policies. It came naturally for the ministry to defend agricultural interests. The existence of a consensus on certain policy principles and procedures to approach policy problems united farmers and the Ministry of Agriculture against outsiders. In Sweden, the agricultural policy network does not privilege the interests of farmers in the same way. Since the early 1960s, there has been no consensus on policy principles and procedures to approach policy problems within the agricultural network. The membership of consumers prevented it from developing such a consensus. Farmers therefore have had difficulties in mobilising political support from agricultural authorities. In nitrate policy making and in the 1990 agricultural reform process, they could not persuade the Agricultural Board and the Agricultural Marketing Board to support them. These findings show that, politically, Swedish farmers are not as privileged as Danish farmers. Thus, policy network structures affect power relations.

Dowding (1995, p. 145) argues that policy networks cannot form 'the centre-piece of explanation.' He is right in that not many network studies are convincing in terms of showing that network structures explain outcomes. However, although network analysts have been most concerned with description, it does not follow that network analysis has no potential for developing into a theory which can explain policy outcomes. This book has shown that one cannot explain agri-environmental policy making and agricultural policy reforms without dealing with the impact of agricultural policy networks. Differences in their degree of cohesion explain why Swedish agricultural authorities had policy preferences which differed from those of EC and Danish agricultural authorities. These differences explain why EC and Danish farmers could form a stronger coalition against outsiders than could their Swedish colleagues. This to a large extent explains why policy choices differed. What is more, the networks had an influence on the strategies of farmers' associations because networks create constraints and opportunities for the pursuit of farmer interests. While the EC and Danish farmers chose a strategy based on confrontation with outsiders, Swedish farmers were more cooperative. These findings clearly show that Dowding fails to see the potentials of network analysis. One *cannot* explain policy outcomes without addressing the influence of network structures.

If the policy network model developed in this book is tested in more policy sectors, and revised if necessary, it may, eventually, achieve status as a theory which can predict policy choices. In the introductory chapter, it was asked whether policy network analysis can help to explain why policy reforms are different and why new, unwelcome policies differ. Basing the answer on the comparison of agri-environmental policy making and agricultural policy reforms, we can conclude that network analysis helps to explain why policies differ. However, to increase the explanatory power of the network model, it was necessary to combine it with an analysis of state structures and of the structural parliamentary power of the organised interests involved in the policy process under scrutiny. The approach applied was exploratory and therefore more research into the relationship between policy networks, state structures and social groups' structural power in parliament is needed. Future research should also attempt to establish the relative explanatory power of each of these variables.

The pluralist inspired group interaction approach within the network literature focuses on process rather than on structures when attempting to explain policy outcomes. Although policy choices are made in a process in which political agents interact, process is only a secondary variable. Policy processes are structured by policy networks. Network structures favour certain options and preclude others. For example, the case studies demonstrated that questioning the idea of subsidising farmers was much more difficult in the European Community than in Sweden because the EC agricultural policy network excluded the question from the political agenda. By contrast, the consumers' central position in the Swedish network and the lack of consensus on the principle of state responsibility enabled reformers to question the whole idea of subsidising farmers. The comparison of agricultural reform processes showed that in the EC, deregulation was not even on the agenda. The question was not whether or not the EC should subsidise agriculture, but how it should be subsidised. In Sweden, there were intense debates about whether the state should continue to subsidise farmers or not. It is within constraints like these the policy process takes place. We must, therefore, take networks seriously. They bias the policy process in certain directions. The limitations and opportunities embodied in a network affect to a large extent the outcomes of the policy process, but they rarely fully determine the specific choice of policy. Even the tightest and most closed policy networks have room for manoeuvre.

Policy instrument analysts should also take policy networks and the broader political context within which they are embedded into account. Too often, they do not consider how meso-level policy networks favour the choice of certain policy instruments and preclude the choice of others. What is possible in one policy sector may not be possible in another, because established sectoral, or sub-sectoral, policy networks differ. Moreover, the choice of certain instruments in a specific sector of one country may not be applicable in a similar sector of another country since the established networks provide different opportunities for political actors to advance or prevent the choice of certain instruments. For instance, this book showed that while the Swedes could apply a fertiliser tax, Danish environmental interests did not succeed in achieving sufficient political support for such an instrument although they tried hard. In Denmark, the character of the agricultural policy network privileged the interests of farmers in opposing green taxes whereas Swedish farmers could not derive similar political power from the agricultural policy network. These findings also show that experiences in one country may not be relevant in other countries.

190

Whether experiences directly can be transferred depends on the character of policy networks. Transfer of policies which run counter to crucial interests in a policy sector seems easiest when no cohesive network exists. The existence of an issue network may provide the best opportunities for policy transfer.

The findings of this book also have implications for policy designers. Policy solutions which are efficient in terms of social welfare may not be 'efficient' in terms of political feasibility. Although a certain solution is efficient for society as a whole, powerful political actors need not perceive it as an efficient solution for them. An important source of their power is the presence of meso-level policy networks which enable them to oppose policy options which they believe are disadvantageous. Therefore, policy makers are not likely to adopt a solution which is efficient for society as a whole if it is opposed by an interest group which is a member of a cohesive network in the policy sector in which the solution could be employed. To design policies that are politically feasible, policy designers need to consider the impact of policy networks.

Policy network analysis is a fruitful approach. Although its emphasis has been on description, this should not lead one to conclude that it has no potential for providing explanation. What is needed in future network studies is an emphasis on explanation rather than description. Most studies applying the network approach are single case studies. Such a research method is less suitable for building theory because it involves the risk that the researcher reduces network structures to actor preferences and thus overlooks the influence of the networks themselves. This book demonstrates the advantages of longitudinal comparative case studies in establishing the consequences of policy networks. Therefore, to progress, policy network analysts should apply such a method.

Bibliography

Aarhus Stifttidende, 20 June 1993.

Administrationsdepartementet (1987), *Turnusgennemgangen af Landbrugsministe-riet. Delrapport 3*, Copenhagen.

AE, (Agra Europe), various issues.

Agra Europe Special Report no. 65 (1992), *The CAP Reform Agreement: An Analysis and Interpretation*, Agra Europe: London.

Akademiet for de Tekniske Videnskaber (1990), *Vandmiljøplanens tilblivelse og iværksættelse*, ATV: Lyngby.

Almond, Gabrial A. (1983), 'Corporatism, Pluralism and Professionel Memory', *World Politics*, vol. 35, no. 2, 1982-83, pp. 245-60.

Almond, Gabrial A. and Powell, G. Bingham (1966), *Comparative Politics: A Developmental Approach*, Little Brown and Company: Boston.

Andersen, Mikael Skou (1994), *Governance by Green Taxes: Making Pollution Prevention Pay*, Manchester University Press: Manchester.

Andersen, Mikael Skou, Christiansen, Peter Munk and Winter Søren (1992), *The Legacy of Environmental Policy in Denmark: Policy, Politics and Implementation*, paper presented at the Workshop in Comparative Research on Environmental Administration and Policy-making, Drøbak, Norway 11-14 June 1992, Department of Political Science, Aarhus University, Aarhus.

Andersen, Mikael Skou and Daugbjerg, Carsten (1994), 'Land-Use Policy in Denmark', in K. Eckerberg, P.K. Mydske, A. Niemi-Iilahti and K.H. Pedersen (eds), *Comparing Nordic and Baltic Countries - Environmental Problems and Policies in Agriculture and Forestry*, TemaNord 1994: 572, Nordic Council of Ministers: Copenhagen, pp. 8-20.

Andersen, Mikael Skou and Hansen, Michael W. (1991), *Vandmiljøplanen: Fra forhandling til symbol*, Niche: Harlev J.

Atkin, Michael (1993), *Snouts in the Trough: European Farmers, the Common Agricultural Policy and the Public Purse*, Woodhead Publishing: Cambridge.

Atkinson, Michael M. and Coleman, William D. (1985), 'Corporatism and Industrial Policy', in Alan Cawson (ed.), *Organized Interests and the State: Studies in*

192

Meso-Corporatism, Sage: London, pp. 22-44.

Atkinson, Michael M. and Coleman, William D. (1989), 'Strong States and Weak States: Sectoral Policy Networks in Advanced Capitalist Economies', *British Journal of Political Science*, vol. 19, pp. 47-67.

Atkinson, Michael M. and Coleman, William D. (1992), 'Policy Networks, Policy Communities and the Problem of Governance' *Governance*, vol. 5, no. 2, pp. 154-80.

Averyt, William F. jr. (1977), *Agropolitics in the European Community: Interest Groups and the Common Agricultural Policy*, Praeger Publishers: New York.

Bachrach, Peter and Baratz, Morton S. (1962), 'Two Faces of Power', *American Political Science Review*, vol. 56, no. 4, pp. 947-52.

Bager, Torben and Søgaard, Villy (1994), *Landmanden og miljøet - holdninger og adfærd belyst ved en spørgeskemaundersøgelse*, Sydjysk Universitetsforlag: Esbjerg.

Baldock, David (1991), 'Introduction', in David Baldock and Graham Bennett (eds), *Agriculture and the Polluter Pays Principle: A Study of Six EC Countries*, Institute for European Environmental Policy: Arnhem and London, pp. 13-14.

Baldock, David (1991a), 'The Polluter Pays Principle and Agriculture', in David Baldock and Graham Bennett (eds), *Agriculture and the Polluter Pays Principle: A Study of Six EC Countries*, Institute for European Environmental Policy: Arnhem and London, pp. 15-31.

Baumgartner, Frank R. and Jones, Bryan D. (1993), *Agendas and Instability in American Politics*, The University of Chicago Press: Chicago.

Bekendtgørelse no. 568 (1988) 'Bekendtgørelse om husdyrgødning og ensilage m.v.', Ministry of the Environment, Copenhagen.

Bekendtgørelse no. 1121 (1992) 'Bekendtgørelse om erhvervsmæssigt dyrehold, husdyrgødning og ensilage m.v.', Ministry of the Environment, Copenhagen.

Bennett, Graham and Baldock, David (1991), 'Conclusions', in David Baldock and Graham Bennett (eds), *Agriculture and the Polluter Pays Principle: A Study of Six EC Countries*, Institute for European Environmental Policy: Arnhem and London, pp. 221-31.

Benson, J.K. (1982), 'A Framework for Policy Analysis', in D. Rogers, D. Whitten and associates (eds), *Interorganizational Coordination*, Iowa State University Press: Ames, pp. 137-76.

Berlingske Tidende, 31 October 1988.

Betænkning no. 561 (1970), *Betænkning afgivet af udvalget vedrørende landbrugsordningerne*, Statens Trykningskontor: Copenhagen.

Betænkning no. 795 (1977), *En fremtidig landbrugspolitik: Betænkning fra Udvalget for den fremtidige landbrugspolitik*, Statens Trykningskontor: Copenhagen.

Betænkning no. 993 (1983), *En fremtidig landbrugspolitik - nogle mere langsigtede perspektiver: Betænkning fra Udvalget vedrørende landbrugets økonomiske vilkår og udvikling*, Direktoratet for Statens Indkøb: Copenhagen.

Betænkning no. 1078 (1986), *Landbokommissionen, 2. delbetænkning: Landbrug og miljø*, Landbrugsministeriet: Copenhagen.

BEUC (1992) 'A Step in the Right Direction: BEUC Opinion on the 1992 Reforms of the Common Agricultural Policy', BEUC/276/92, 6 November 1992, Brussels.

Birgersson, Bengt Owe and Westerståhl, Jörgen (1992), *Den svenska folkstyrelsen* (5ᵗʰ ed.), Publica: Stockholm.

Blondel, Jean (1968), 'Party Systems and Patterns of Government in Western Europe' *Canadian Journal of Political Science* vol. 1, no. 2, pp. 180-203.

Bolin, Olof, Meyerson, Per-Martin, and Ståhl, Ingemar (1984), *Makten över maten: Livsmedelsektorens politiska ekonomi*, Studieförbundet Näringsliv och Samhälle, SN&S: Stockholm. (Also published in English in 1986, titled *Political Economy of the Food Sector: The Case of Sweden*, SNS Förlag, Stockholm).

Brand, Jack (1992), *British Parliamentary Parties: Policy and Power*, Clarendon Press: Oxford.

Bregnsbo, Henning and Niels Chr. Sidenius (1993). 'Denmark: The National Lobby Orchestra', in M.C.P.M. van Schendelen (ed.), *National Public and Private EC Lobbying*, Dartmouth: Aldershot, pp. 183-200.

Bressers, Hans Th. A. (1995), *Policy Networks and the Choice of Instruments*, NIG working paper no. 95-6, Netherlands Institute of Government: Enschede.

Bretherton, Charlotte and Sperling, Liz (1996), 'Women's Networks and the European Union: Towards an Inclusive Approach', *Journal of Common Market Studies*, vol. 34, no. 4, pp. 487-508.

Brewin, Christopher (1987), 'The European Community: a Union of States without Unity of Government', *Journal of Common Market Studies*, vol. 26, no. 1, pp. 1-23.

Bryld, Claus (1992), *Den demokratiske socialismes gennembrudsår: Studier i udformningen af arbejderbevægelsens politiske ideologi i Danmark 1884-1916 på den nationale og internationale baggrund*, Selskabet til forskning i arbejderbevægelsens historie: Copenhagen.

Buchanan, David, et.al. (1988), 'Getting In, Getting On, Getting Out, and Getting Back', in Alan Bryman (ed.), *Doing Research in Organisations*, Routledge: London, pp. 53-67.

Buksti, Jacob A. (1974), *Et enigt landbrug? Konflikt og samarbejde mellem landbrugets organisationer: En studie i landbrugets organisationsforhold med udgangspunkt i perioden 1957-61*, Erhvervsarkivet og Universitetsforlaget i Aarhus: Aarhus.

Buksti, Jacob A. (1980), 'Udviklingen i landbrugets organisationsforhold 1972-79', in Jacob A. Buksti (ed.), *Organisationer under forandring: Studier i organisationssystemet i Danmark*, Forlaget Politica: Aarhus, pp. 283-308.

Buksti, Jacob A. (1983), 'Bread-and-Butter Agreement and High Politics Disagreement: Some Reflections on the Contextual Impact on the Agricultural Interests in EC Policy-Making', *Scandinavian Political Studies*, vol. 6, no. 4, pp. 261-80.

Buksti, Jacob A. (1983a), 'Organisationer og offentlig politik: Interesseorganisationernes deltagelse i den politiske beslutningsproces i Danmark og den samfundsmæssige udvikling på området', *Nordisk Administrativt Tidsskrift*, vol. 64, no. 2, pp. 191-212.

Buksti, Jacob A. and Johansen, Lars Nørby (1979), 'Varations in Organizational Participation in Government: The Case of Denmark', *Scandinavian Political Studies*, vol. 2, no. 3, pp. 197-220.

Burkhardt-Reich, B. and Schumann, W. (1983), *Agrarverbande in der EG*, N.P. Engel Verlag: Kehl am Rein.

Butt, Philip Alan (1985), *Pressure Groups in the European Community*, UACES: London

Børsen, 20 September 1985.

Campbell, John. C. with Baskin, Mark A., Baumgartner, Frank R., and Halpern, Nina P. (1989), 'Afterword on Policy Communities: A Framework for Comparative Research' *Governance*, vol. 2, no. 1, pp. 86-94.

Carlsson, Sten (1979), 'Protektionister och frihandlare 1885-95', in Jan Cornell (ed.), *Den svenska historien 13: Emigrationen och det industrielle genombrottet*, Albert Bonniers Förlag: Stockholm, pp. 94-109.

Cavanagh, Michael, Marsh, David and Smith, Martin (1995), 'The Relationship between Policy Networks at the Sectoral and Sub-sectoral Levels: A Response to Jordan, Maloney and McLaughlin', *Public Administration* vol. 43, no. 4, pp. 627-29.

Cawson, Alan (1985), 'Introduction. Varieties of Corporatism: the Importance of the Meso-level of Interest Intermediation', in Alan Cawson (ed.), *Organized Interests and the State: Studies in Meso-Corporatism*, Sage: London, pp. 1-21.

Cerny, Philip G. (1990), *The Changing Architecture of Politics: Structure, Agency and the Future of the State*, Sage: London.

Christensen, Dag Arne (1994), 'Fornyinga av bondepartia i Noreg og Sverige', in Knut Heider and Lars Svåsand (eds), *Partiene i en brytningstid*, Alma Mater Forlag: Bergen, pp. 327-54.

Christensen, Jørgen Grønnegaard (1978), 'Da centraladministrationen blev international: En analyse af den administrative tilpasningsproces ved Danmarks tilslutning til EF', in Niels Amstrup and Ib Faurby (eds), *Studier i dansk udenrigspolitik*, Forlaget Politica: Aarhus, pp. 75-118.

Christensen, Jørgen Grønnegaard (1980), *Centraladministrationen: Organisation og politisk placering*, Jurist- og Økonomforbundets Forlag: Copenhagen.

Christensen, Jørgen Grønnegård (1981), 'Blurring the International-Domestic Politics Distinction: Danish Representation at EC Negotiations', *Scandinavian Political Studies*, vol. 4, no. 3, pp. 191-208.

Christensen, Jørgen Grønnegaard (1985), 'In search of unity: cabinet committees in Denmark', in Thomas T. Mackie and Brian Hogwood (eds), *Unlocking the Cabinet: Cabinet Structures in Comparative Perspective*, Sage: London, pp. 114-37.

Christensen, Jørgen Grønnegaard (1987), 'Hvem har magten over miljøpolitikken: Politikerne, embedsmændene eller organisationerne', in Alex Dubgaard (ed.), *Relationer mellem landbrug og samfund: En foredragssamling*, report no. 36, Statens Jordbrugsøkonomiske Institut: Copenhagen.

Christensen, Jørgen Grønnegård and Christiansen, Peter Munk (1992), *Forvaltning og Omgivelser*, Systime: Herning.

Christensen, Jørgen Grønnegaard and Ibsen, Marius (1991), *Bureaukrati og bureaukrater*, Systime: Herning.

Christensen, Tom and Egebjerg, Morten (1979), 'Organized Group-Government Relations in Norway: On the Structured Selection of Participants, Problems, Solutions and Opportunities', *Scandinavian Political Studies*, vol. 2, no. 3, pp. 239-59.

Christiansen, Peter Munk (1990), 'Udgiftspolitikken i 1980'erne: Fra asymmetri til asymmetrisk tilpasning?', *Politica*, vol. 22, no. 4, pp. 442-56.

Christiansen, Peter Munk (1994), 'A Negotiated Economy? Public Regulation of the Manufactoring Sector in Denmark', *Scandinavian Political Studies*, vol. 17, no. 4, pp. 305-19.

Christiansen, Peter Munk (1996), 'Denmark', in Peter Munk Christiansen (ed.), *Governing the Environment: Politics, Policy and Organization in the Nordic Countries*, Nord 1996:5, Nordic Council of Ministers: Copenhagen, pp. 29-101.

Cloke, Paul and Le Heron, Richard (1994), 'Agricultural Deregulation: The Case of New Zealand', in Philip Lowe, Terry Marsden and Sarah Whatmore (eds), *Regulating Agriculture*, David Fulton Publishers: London, pp. 104-26.

Collins, Neil (1990), 'The European Community's Farm Lobby', *Corruption and Reform*, vol. 5, no. 3, pp. 235-57.

Commission of the European Communities (1958), *Recueil des documents de la Conference Agricole des Etats Membres de la Communauté Economique Européenne a Stresa du 3 au 12 juillet 1958*, Brussels.

COPA (1986), 'The Committee of Agricultural Organisations in the European Community, C.O.P.A.', no. 2536, Brussels 16 Dec. 1986.

COPA (1991), 'COPA Proposal on the Future of the Common Agricultural Policy (CAP)', Brussels, 23 May 1991, Pr (91) 13.

COPA (1992) 'Press Release: Politicians Should Take Full Responsibility for the Decisions on the Common Agricultural Policy Reform', CdP (92) 20, 22 May 1992, Brussels.

Cunningham, Caroline (1992), 'Sea Defences: A Professionalized Network', in David Marsh and R.A.W. Rhodes (eds), *Policy Networks in British Government*, Clarendon Press: Oxford, pp. 100-23.

Dahl, Ottar (1980), *Grunntrekk i historieforskningens metodelære* (2nd ed.), Universitetsforlaget: Oslo.

Dahl, Robert A. (1969), 'The Concept of Power', in Roderick Bell, David Edwards and R. Harrison Wagner (eds), *Political Power: A Reader in Theory and Research*, The Free Press/Collier-Macmillan: New York/London, pp. 79-93.

Damgaard, Erik (1981), 'Politiske sektorer: Jerntrekanter eller løse netværk?', *Nordisk administrativt tidsskrift*, vol. 62, no. 4, pp. 396-411.

Damgaard, Erik (1982), *Partigrupper, repræsentation og styring*, Schultz Forlag: Copenhagen.

Damgaard, Erik (1982a), 'The Public Sector in a Democratic Order: Problems and

Non-Solutions in the Danish Case', *Scandinavian Political Studies*, vol. 5, no. 4, pp. 337-58.

Damgaard, Erik (1984), 'The Importance and Limits of Party Government: Problems of Governance in Denmark', *Scandinavian Political Studies*, vol. 7, no. 2, pp. 97-110.

Damgaard, Erik (1986), 'Causes, Forms, and Consequences of Sectoral Policy-Making: Some Danish Evidence', *European Journal of Political Research*, vol. 14, no. 3, pp. 273-87.

Damgaard, Erik (1992), 'Denmark: Experiments in Parliamentary Government', in Erik Damgaard (ed.), *Parliamentary Change in the Nordic Countries*, Scandinavien University Press: Oslo.

Damgaard, Erik (1994), 'The Strong Parliaments of Scandinavia: Continuity and Change of Scandinavian Parliaments', in Gary W. Copeland and Samual C. Patterson (eds), *Parliaments in the Modern World: Changing Institutions*, The University of Michigan Press: Ann Arbor, pp. 85-103.

Damgaard, Erik and Eliassen, Kjell A. (1978), 'Corporate Pluralism in Danish Law-Making', *Scandinavian Political Studies*, vol. 1, no. 4, pp. 285-313.

Danmarks Naturfredningsforening (1991), *Bæredygtigt landbrug*, Copenhagen.

Danske Mejeriers Mælkeudvalg (1992), *Otte år med mælkekvoteordningen 1984-92: Hovedrapport*, Aarhus.

Daugbjerg, Carsten (1992), *Policy Networks, Regulation and Danish Agricultural and Industrial Organised Interests*, Department of Political Science Aarhus University, Aarhus (unpublished MA dissertation).

Daugbjerg, Carsten (1993), *A Preliminary Model for the Study of Agriculture and the Environment in Scandinavia*, paper presented at the Nordic Political Science Congress in Oslo, 19-21 August 1993, Department of Political Science Aarhus University, Aarhus.

Daugbjerg, Carsten (1994), 'Dansk industri og det indre marked: Policy-netværk i Bruxelles og København', *Politica*, vol. 26, no. 4, pp. 456-73.

Daugbjerg, Carsten (1994a), *Policy Networks and Changing Environments. Environmental Regulation and Agricultural Reform: Some Theoretical Considerations*, Department of Political Science, Aarhus University, Aarhus.

Daugbjerg, Carsten (1994b), 'Den danske landbrugssektor 1993: Brancheprofil', in Mikael Skou Andersen (ed.), *Spredning af renere teknologi i landbruget*, Arbejdsrapport nr. 59, Miljøstyrelsen: Copenhagen, pp. 79-103.

Daugbjerg, Carsten (1995), 'Er miljøpolitik teknik eller politik? En komparativ analyse af miljøpolitik i den svenske og danske landbrugssektor', *Nordisk Administrativt Tidsskrift*, vol. 76, no. 1, pp. 33-47.

Daugbjerg Carsten (1995a), 'Instruments, Organisation and Goals of Environmental Policy: Typologies and Classifications', paper to 'Nordiskt forskarsymposium om miljö och samhälle', Gothenburg, Sweden, 13-15 June, 1995.

Daugbjerg Carsten (1995b), *Policy Network and Policy Choices in Changing Environments: Environmental Policy in Swedish and Danish Agriculture*, paper to the Political Studies Asociation Annual Conference, York University, 18-20

April, 1995, Department of Political Science, Aarhus University, Aarhus.

Daugbjerg, Carsten (1996), 'Landbrug og miljø i Danmark og Sverige: Forskellige løsninger på samme problem', *Miljøforskning* no. 25, October 1996.

Daugbjerg, Carsten (1996a), *Policy Networks and Agricultural Policy Reforms in Sweden and the European Community: Explaining Different Policy Outcomes*, paper to the Political Science Association Annual Conference, University of Glasgow, 10-12 April, 1996, Department of Political Science, Aarhus University, Aarhus.

Daugbjerg, Carsten (1997), 'Policy Networks and Agricultural Policy Reforms: Explaining Deregulation in Sweden and Re-Regulation in the European Community', *Governance*, vol. 10, no. 2, 123-41.

Daugbjerg, Carsten (1998), 'Similar Problems, Different Policies: Policy Networks and Environmental Policy in Swedish and Danish Agriculture', in David Marsh (ed.), *Policy Networks: Theory and Comparison* (provisional title), Open University Press: Milton Keynes (forthcoming).

De danske Landboforeninger and Danske Husmandsforeninger (1989), *Redegørelse for landbrugets miljøindsats og effekten heraf*, Copenhagen.

De danske Landboforeninger and Danske Husmandsforeninger (1990), *Miljøindsatsen i Landbruget 1990*, Copenhagen.

Den danske Regering (1988), *Regeringens handlingsplan for miljø og udvikling: Opfølgning af anbefalinger fra Verdenskommissionen om miljø og udvikling og i FN's Miljøperspektiv til år 2000*, Statens Informationstjeneste: Copenhagen.

De europæiske Fællesskaber (1987), *Traktater om oprettelse af De europæiske Fællesskaber*, Kontoret for De europæiske Fællesskabers officielle Publikationer: Luxembourg.

Det Økonomiske Råds Formandskab (1968), *Dansk økonomi i efteråret 1968*, Copenhagen.

Doern, G. Bruce and Phidd, R. W. (1983), *Canadian Public Policy: Ideas, Structure, Process*, Methuen: Toronto.

Doern, G. Bruce and Wilson, V. Seymour (1974), 'Conclusions and Observations', in G. Bruce Doern and V. Seymour Wilson (eds), *Issues in Canadian Public Policy*, Macmillan of Canada: Toronto, pp. 337-45.

Dogan, Mattei and Pellassy, Dominique (1984), *How to Compare Nations: Strategies in Comparative Politics*, Chatham House Publishers: Chatham.

Dowding, Keith (1995), 'Model or Metaphor? A Critical Review of the Policy Network Approach', *Political Studies*, vol. 43, no. 1, pp. 136-58.

Dubgaard, Alex (1991), *The Danish Nitrate Policy in the 1980s*, report no. 59, Statens Jordbrugsøkonomiske Institut: Copenhagen.

Döhler, Marion (1991), 'Policy Networks, Opportunity Structures and Neo-Conservative Reform Strategies in Health Policy', in Bernd Marin and Renate Mayntz (eds), *Policy Networks: Empirical Evidence and Theoretical Considerations*, Campus Verlag/Westview Press: Frankfurt am Main/Boulder, pp. 235-96.

Easton, David (1953), *The Political System: An Inquiry into the State of Political Science*, Alfred A. Knopf: New York.

Eckerberg, Katarina (1994), 'Consensus, Conflict or Compromise: The Swedish Case', in K. Eckerberg, P.K. Mydske, A. Niemi-Iilahti and K.H. Pedersen (eds), *Comparing Nordic and Baltic Countries - Environmental Problems and Policies in Agriculture and Forestry*, TemaNord 1994: 572, Nordic Council of Ministers: Copenhagen, pp. 76-97.

Eckerberg, K, P.K. Mydske, A. Niemi-Iilahti and K.H. Pedersen (eds) (1994), *Comparing Nordic and Baltic Countries - Environmental Problems and Policies in Agriculture and Forestry*, TemaNord 1994: 572, Nordic Council of Ministers: Copenhagen.

Elmore, Richard F. (1987), 'Instruments and Strategy in Public Policy', *Policy Studies Review*, vol. 7, no. 1, pp. 174-86.

Elvander, Nils (1969), *Intresseorganisationer i dagens Sverige*, (2nd ed.), GWK Gleerup Bokförlag: Lund.

ESC (Economic and Social Committee) (1980), *Community Advisory Committees for the Representation of Socio-Economic Interests*, Saxon House: Farnborough.

Fearne, Andrew (1991), 'The History and the Development of the CAP 1945-1985', in Christopher Ritson and David Harvey (eds), *The Common Agricultural Policy and the World Economy: Essays in Honour of John Ashton*, CAB International: Wallingford, pp. 21-70.

Fearne, Andrew (1991a), 'The CAP Decision-Making Process', in Christopher Ritson and David Harvey (eds), *The Common Agricultural Policy and the World Economy: Essays in Honour of John Ashton*, CAB International: Wallingford, pp. 101-16.

Feldt, Kjell-Olof (1991), *Alla dessa dagar ... I regeringen 1982-1990*, Norstedts Förlag: Stockholm.

Finansdepartementet (1987), *Vägar ut ur jordbruksprisregleringen - några idéskisser: Bidrag till ett seminarium anordnat av expertgruppen för studier i offentlig ekonomi*, Ds Fi 1987: 4, Allmänna Förlaget: Stockholm.

Finansdepartementet (1988), *Alternativ i jordbrukspolitiken: Rapport till expertgruppen för studier i offentlig ekonomi*, Ds 1988: 54, Allmänna Förlaget: Stockholm.

FOA (1988), (Försvarets Forskningsanstalt), *Säkerhetsaspekter på livsmedelförsörjningen*, FOA rapport C 10311-1.2, Stockholm.

Folketingets miljøudvalg, (1973) 'Betænkning over forslag til lov om miljøbeskyttelse', *Folketingstidende*, 124. årg., 1972-73, tillæg B, bind 2.

Folketingstidende C (1984-85), 'Folketingsbeslutning om nedbringelse af forurening med næringssalte og organisk stof', Copenhagen.

Folketingstidende F (1984-85), 'Miljøministerens redegørelse af 12/12 84 vedrørende handlingsplanen for nedbringelse af forurening med næringssalte og organisk stof (Redegørelse nr. R7)', Copenhagen.

Folketingstidende F (1989-90), 'Redegørelse af 8/5 90 om vandmiljøplanen (redegørelse nr. R 17)', Copenhagen.

Forureningsrådet, (1971), *Vand: Målsætning: En redegørelse fra målsætningsudvalget og hovedvandsudvalget*, publikation nr. 15, Statens Trykningskontor: Copenhagen

Froman, Lewis A. jr. (1968), 'The Categorization of Policy Contents', in Austin Ranney (ed.), *Political Science and Public Policy*, Markham Publishing Company: Chicago.

Gad, Holger (1969), 'Mål og midler i dansk landbrugspolitik: Prisstøtte eller indkomststøtte' *Økonomi og Politik*, vol. 43, no. 1, pp. 3-19.

Gardner, Brian (1987), 'The Common Agricultural Policy: The Political Obstacle to Reform', *The Political Quarterly*, vol. 58, pp. 167-179.

Gibbons, J. (1995), 'Public Policy Performance in Spain: The Case of the Farming and Food Sector', paper presented to the Political Studies Association Conference, York University, 18-20 April 1995.

Gibbons, John Patrick (1990), *The Origins and Influence of the Irish Farmers' Association*, Department of Government, Faculty of Economic and Social Sciences, University of Manchester (unpublished PhD-thesis).

Glans, Ingemar (1989), 'Langtidsudviklingen i dansk vælgeradfærd', in Jørgen Elklit and Ole Tonsgaard (eds), *To folketingsvalg. Vælgerholdninger og vælgeradfærd i 1987 og 1988*, Politica: Aarhus.

Glasbergen, Pieter (1992), 'Agro-Environmental Policy: Trapped in an Iron Law?', *Sociologia Ruralis*, vol. 32, no. 1, pp. 30-48.

Graham, George J. (1971), *Methodological Foundations for Policy Analysis*, Xerox College Publishing: Waltham.

Grant, Wyn (1985), 'Introduction', in Wyn Grant (ed.), *The Political Economy of Corporatism*, Macmillan: London, pp 1-31.

Grant, Wyn (1985a) (ed.), *The Political Economy of Corporatism*, Macmillan: London.

Grant, Wyn (1991), *The Dairy Industry: An International Comparison*, Dartmouth: Aldershot.

Grant, Wyn (1992), 'Models of interest intermediation and policy formation applied to an internationally comparative study of the dairy industry', *European Journal of Political Research*, vol. 21, no. 1-2, pp. 53-68.

Grant, Wyn (1993) *Business and Politics in Britain* (2nd ed.), Macmillan: London.

Grant, Wyn (1993a). 'Pressure Groups and the European Community: An Overview', in Sonia Mazey and Jeremy J. Richardson (eds), *Lobbying in the European Community*, Oxford University Press: Oxford, pp. 27-46.

Grant, Wyn (1995), 'The limits of Common Agricultural Policy reform and the option of denationalization', *Journal of European Public Policy*, vol. 2, no. 1, pp. 1-18.

Grant, Wyn (1997), *The Common Agricultural Policy*, Macmillan: London.

Grant, Wyn, Paterson, William, and Whitson, Colin (1988), *Government and the Chemical Industry: A Comparative Study of Britain and West Germany*, Clarendon Press: Oxford.

Gray, O.W. (1989), *Pressure Groups and their Influence on Agricultural Policy and its Reform in the European Community*, University of Bath, (unpublished PhD-thesis).

Guyomard, H., Mahé, L.P., Munk, K.J and Roe, T.L (1993), 'Agriculture in the Uruguay Round: Ambitions and Realities' *Journal of Agricultural Economics*, vol. 44, no. 2, pp. 245-63.

Hall, John A. and Ikenberry, G. John (1989), *The State*, Open University Press: Milton Keynes.

Hall, Peter A. (1986), *Governing the Economy: The Politics of State Intervention in Britain and France*, Polity Press: Cambridge.

Hall, Peter (1993), 'Policy Paradigms, Social Learning, and the State: The Case of Economic Policymaking in Britain', *Comparative Politics*, vol. 25, no. 3, pp. 275-96.

Hall, Peter A. and Taylor, Rosemary C.R. (1996), 'Political Science and the Three New Institutionalisms', *Political Studies*, vol. 44, no. 5, pp. 936-57.

Hall, Phoebe, Land, Hilary, Parker, Roy and Webb, Adrian (1975), *Change, Choice and Conflict in Social Policy*, Heinemann: London.

Hansen, Henning Otte (1993), *Landbrugets placering i samfundsøkonomien*, Jord-brugsforlaget: Frederiksberg.

Hansen, Karin and Torpe, Lars (1977), *Socialdemokratiet og krisen i 1930'erne*, Modtryk: Aarhus.

Harris, Simon, Swinbank, Alan and Wilkinson, Guy (1983), *The Food and Farm Policies of the European Community*, John Wiley and Sons: Chichester.

Harvey, D.R. (1994), 'Agricultural Policy Reform after the Uruguay Round', in K.A. Ingersent, A.J. Rayner and R.C. Hine (eds), *Agriculture in the Uruguay Round*, St. Martin's Press: New York, pp. 223-59.

Hay, Colin (1995), 'Structure and Agency', in David Marsh and Gerry Stoker (eds), *Theories and Methods in Political Science*, Macmillan: London, pp. 189-206.

Heclo, Hugh (1978), 'Issue Networks and the Executive Establishment', in Anthony King (ed.), *The New American Political System*, American Enterprise Institute for Public Policy Research: Washington D.C, pp. 87-124.

Hedrén, Johan (1993), 'Vetenskap, teknologi och miljöpolitik', paper presented at Nordiska statsvetenskapliga kongressen in Oslo, August 19-21, 1993.

Heisler, Martin O. (1979), 'Corporate Pluralism Revisited: Where is the Theory?', *Scandinavian Political Studies*, vol. 2, no. 3, pp. 277-79.

Heisler, Martin O. and Kvavik, Robert B (1974), 'Patterns of European Politics: the "European Polity" Model', in Martin O. Heisler (ed.), *Politics in Europe: Structures and Processes in Some Post-Industrial Democracies*, McKay: New York.

Hellevik, Ottar (1987), *Forskningsmetode i sosiologi og statsvitenskap* (4th ed.), Universitetsforlaget: Oslo.

Hellström, Gunnar (1976), *Jordbrukspolitik i industrisamhället - med tyngdepunkt på 1920- och 30-talen*, LTs förlag: Stockholm.

Hendriks, Gisela (1991), *Germany and European Integration*, Berg: New York and Oxford.

Hendriks, Gisela (1994), 'German Agricultural Policy Objectives', in Rasmus Kjel-dahl and Michael Tracy (eds) *Renationalisation of the Common Agricultural Policy?* Institute of Agricultural Economics/Agricultural Policy Studies: Valby/la

Hutte, pp. 59-73.

Hill, Berkeley (1993), 'The "Myth" of the Family Farm: Defining the Family Farm and Assessing its Importance in the European Community', *Journal of Rural Studies*, vol. 9, no. 4, pp. 359-70.

Holmberg, Søren and Esaiasson, Peter (1988), *De folkvalda: En bok om riksdagledamöterna och den representativa demokratin i Sverige*, Bonniers: Stockholm.

Howlett, Michael (1991), 'Policy Instruments, Policy Styles, and Policy Implementation: National Approaches to Theories of Instrument Choice', *Policy Studies Journal*, vol. 19, no. 2, pp. 1-21.

Howlett, Michael and Ramesh, M. (1994), 'Patterns of Policy Instrument Choice: Policy Styles, Policy Learning and the Privatization Experience', *Policy Studies Review*, vol. 12, no. 1/2, pp. 3-24.

Husmandsforeninger, Danske (1985), *Beretning 1985*, Copenhagen.

Husmandsforeninger, Danske (1987), 'Vedrørende regeringens handlingsplan mod forurening med næringsstoffer af det danske vandmiljø', Copenhagen, April 1, 1987, published in *Folketingstidende B* 'Bilagshæfte til Beretning over Vandmiljøplanen afgivet af miljø- og planlægningsudvalget den 30. april 1987', pp. 111-13.

Husmandsforeninger, Danske (1988), *Beretning 1988*, Copenhagen.

Husmandsforeninger, Danske (1990), *Beretning 1990*, Copenhagen.

Husmandsforeninger, Danske (1991), *Danske Husmandsforeningers politik på et bæredygtigt landbrug*, Copenhagen.

Husmandsforeninger, Danske (1991a), *Beretning 1991*, Copenhagen.

Husmandsforeninger, Danske (1992), *Beretning 1992*, Copenhagen.

Ingersent, K.A., Rayner, A.J. and Hine, R.C. (1994), 'The EC Perspective', in K.A. Ingersent, A.J. Rayner and R.C. Hine (eds), *Agriculture in the Uruguay Round*, St. Martin's Press: New York, pp. 55-87.

Jansen, Alf-Inge and Lundqvist, Lennart J. (1993), 'Om valet af policyinstrument i miljöpolitiken', paper to the NERP project meeting in Copenhagen, December 1-2, 1993.

Jenkins-Smith, Hank C. and Sabatier, Paul A. (1993), 'The Dynamics of Policy-Oriented Learning', in Paul A. Sabatier and Hank C. Jenkins-Smith (eds), *Policy Change and Learning: An Advocacy Coalition Framework*, Westview Press: Boulder , pp. 41-56.

Jordan, A. Grant (1981), 'Iron Triangles, Wholly Corporatism and Elastic Nets: Images of the Policy Process', *Journal of Public Policy*, vol. 1, no. 1, pp. 95-123.

Jordan, Grant (1984), 'Pluralistic Corporatisms and Corporate Pluralism', *Scandinavian Political Studies*, vol. 7, no. 3, pp. 137-53.

Jordan, Grant (1990), 'Policy Community Realism Versus "New" Institutionalist Ambiguity', *Political Studies*, vol. 38, no. 3, pp. 470-84.

Jordan, Grant (1990a), 'Subgovernments, Policy Communities and Networks. Refilling the old Bottles', *Journal of Theoretical Politics*, vol. 2, no. 3, pp. 319-38.

Jordan, Grant (1990b), 'The Pluralism of Pluralism: An Anti-theory?', *Political Studies*, vol. 38, no. 2., 1990, pp. 286-301.

Jordan, Grant, Maloney, William A. and McLaughlin, Andrew M. (1994), 'Characterizing Agricultural Policy-Making', *Public Administration*, vol. 72, no. 4, pp. 505-26.

Jordan, Grant and Maloney, William (1995), 'Re-establishing the Micro-level Credentials of the Public Policy Community Idea', Department of Government, University of Strathclyde.

Jordan, Grant and Maloney, William (1995a), 'Policy Networks Expanded: A Comment on Cavanagh, Marsh and Smith', *Public Administration* vol. 43, no. 4, pp. 630-3.

Jordan, A. G. and Richardson, J. (1982), 'The British Policy Style or the Logic of Negotiation', in Jeremy Richardson (ed.), *Policy Styles in Western Europe*, George Allan & Unwin: London.

Jordan, A. G. and Richardson, J.J. (1987), *Government and Pressure Groups in Britain*, Clarendon Press: Oxford.

Jordan, A.G and Richardson, J.J. (1987a), *British Politics and the Policy Process: An Arena Approach*, Allen & Unwin: London.

Jordan, Grant and Schubert, Klaus (1992), 'A Preliminary Ordering of Policy Network Labels', *European Journal of Political Research*, vol. 21, no. 1-2, pp. 7-27.

Jordbruksdepartementet (1986), *Åtgärder för at minska spannmålsproduktionen på kort sikt. PM av spannmålsgruppen*, Ds Jo 1986: 2, Allmänna Förlaget: Stockholm.

Jordbruksdepartementet (1987), *Intensiteten i jordbruksproduktionen: Miljöpåverkan och spannmålsöverskott: Betänkande av arbetsgruppen med uppgift att utreda vissa frågor rörande en lägre intensitet i jordbruksproduktionen*, Ds Jo 1987: 3, Allmänna Förlaget: Stockholm.

Jordbruksdepartementet (1987a), *Åtgärder för at minska spannmålsöverskottet och stimulera alternativ markanvänding: Rapport 3 från spannmålsgruppen*, Ds Jo 1987: 2, Allmänna Förlaget: Stockholm.

Jordbruksdepartementet (1989), *En ny livsmedelpolitik*, Ds 1989: 63, Allmänna Förlaget: Stockholm.

Jordbrukdepartementet (1990), *Sammanställning av remissytrandena över rapporten säkerhetspolitiska aspekter på livsmedelsförsörjningen*, Ds 1990: 21, Allmänna Förlaget: Stockholm.

Jordbrukdepartementet (1990a), *Alternativ i jordbrukspolitiken samt arealbidrag: Sammanställninger av remissyttrandena*, Ds 1990: 25, Allmänna Förlaget: Stockholm.

Jordbruksverket (1992), *Miljöavgifter. Bekämpningsmedel, handelsgødsel*, Report 1992: 41, Jönköping.

JoU 1976/77: 26, 'Jordbruksutskottets betänkande 1976/77:26 med anledning av motioner om vatten och luftvård, m.m., Stockholm.

JoU 1978/79: 16, 'Jordbruksutskottets betänkande 1978/79: 16 med anledning av motioner om använding av handelsgödsel m.m. i jord- och skogsbruket', Stockholm.

JoU 1983/84: 15, 'Jordbruksutskottets betänkande 1983/84: 15 om vissa ekonomiska åtgärder m.m.', Stockholm.

JoU 1983/84: 20, 'Jordbruksutskottets betänkande 1983/84: 20 om vissa livsmedel-politiska frågor (prop. 1983/84: 76)', Stockholm.

JoU 1984/85: 33, 'Jordbruksutskottets betänkande 1984/85: 33 om livsmedelpoli-tiken', Stockholm.

JoU 1987/88: 24, 'Jordbruksutskottets betänkande 1987/88: 24 om miljöförbättrande åtgärder i jordbruket, m.m. (prop. 1987/88: 128)', Stockholm.

JoU 1989/90: 25, 'Jordbruksutskottets betänkande 1989/90: 25. Livsmedelspoliti-ken', Stockholm.

JoU 1990/91: 30, 'Jordbruksutskottets betänkande 1990/91: 30. Miljöpolitikken', Stockholm.

Judge, David (1990), Parliament and Industry, Dartmouth: Aldershot.

Judge, David (1993), The Parliamentary State, Sage: London.

Just, Flemming (1992), Landbruget, staten og eksporten 1930-1950, Sydjysk Uni-versitetsforlag: Esbjerg.

Just, Flemming (1994), 'Agriculture and Corporatism in Scandinavia', in Philip Lowe, Terry Marsden, and Sarah Whatmore (eds), Agricultural Regulations, Da-vid Fulton Publishers: London, pp. 31-52.

Just, Flemming (1995), 'Fra Axelborg til Slotsholmen: Forholdet mellem landbrug og stat' Samfundsøkonomen, no. 2, pp. 11-18.

Just, Flemming and Olesen, Thorsten B. (1995), 'Danish Agriculture and the Euro-pean Market Schism 1945-1960', in Thorsten B. Olsen (ed.), Interdependence Versus Integration: Denmark, Scandinavia and Western Europe, 1945-1960, Odense University Press: Odense, pp. 129-46.

Just, Flemming and Omholt, Knut (1984), Samspillet mellem staten, landbrugsorga-nisationerne og landbrugskooperationen, Sydjysk Universitetsforlag: Esbjerg.

Jyllands-Posten, 24 September 1994, 16 May 1996.

Jönson, Sture (1989), 'Lantbrukets miljömål - en renere fremtid', in Miljö- och Energidepartementet (ed.), Jordbruket och miljön: Rapport från miljövårdsbe-redningen, Ds 1989: 51, Allmänna Förlaget: Stockholm, pp. 32-36.

Kvale, Steinar (1983), 'The Qualitative Research Interview: A Phenomenological and Hermeneutical Mode of Understanding', Journal of Phenomenological Phy-schology, vol. 14, no. 2, pp. 171-96.

Kvale, Steinar (1984), 'Om tolkningen af kvalitative forskningsinterview', Tidskrift för Nordisk Förening för Forskning, vol. 4, no. 3-4, pp. 55-66.

Kvale, Steinar (1989), 'To Validate is to Question', in Steinar Kvale (ed.), Issues of Validity in Qualitative Research, Studentlitteratur: Lund, pp. 73-92.

Kampmann, Jens, (1980), 'Landbrugets indflydelse på miljøet', in Landbrugsraadet and Miljøministeriet (eds), Landbrug og miljø: Konferencegrundlag og indlæg på landbrugsorganisationernes og miljøministeriets konference i Rønne, septem-ber 1980, Landbrugsraadet: Copenhagen.

Kassim, Hussein (1994), 'Policy Networks, Networks and European Union Policy Making: A Sceptical Review', West European Politics, vol. 17, no. 4, pp. 15-27.

Katzenstein, Peter J. (1978), 'Conclusion: Domestic Structures and Strategies of Foreign Economic Policy', in Peter J. Katzenstein (ed.) *Between Power and Plenty: Foreign Economic Policies of Advanced Industrial States*, The University of Wisconsin Press: Madison.

Keeler, John T.S. (1987), *Farmers, the State, and Agricultural Policy-making in the Fifth Republic*, Oxford University Press: Oxford.

Keeler, John T.S. (1996), 'Agricultural Power in the European Community: Explaining the Fate of CAP and GATT Negotiations', *Comparative Politics*, vol. 28, no. 2, January, pp. 127-49.

Kenis, Patrick and Schneider, Volker (1991), 'Policy Networks and Policy Analysis: Scrutinizing a New AnaLytical Toolbox', in Bernd Marin and Renate Mayntz (eds), *Policy Networks: Emprical Evidence and Theoretical Considerations*, Campus Verlag/Westview Press: Frankfurt am Main/Boulder, pp. 25-59.

Keohane, Robert O. and Hoffmann, Stanley (1991), 'Institutional Change in Europe in the 1980s', in Robert O. Keohane and Stanley Hoffmann (eds), *The New European Community: Decisionmaking and Institutional Change*, Westview Press: Boulder, pp. 1-39.

Kingdon, John W. (1984), *Agendas, Alternatives, and Public Policies*, Harper Collins Publishers: New York.

Kirchheimer, Otto (1966), 'The Transformation of the Western European Party Systems', in Joseph LaPalombara and Myron Weiner (eds), *Political Parties and Political Development*, Princeton University Press: Princeton, pp. 177-200.

Kjeldahl, Rasmus (1994), 'Reforming the Reform? The CAP at a Watershed', in Rasmus Kjeldahl and Michael Tracy (eds), *Renationalisation of the Common Agricultural Policy*, Institute of Agricultural Economics/Agricultural Policy Studies: Valby/la Hutte, pp. 5-22.

Kjeldsen-Kragh, S. (1986), 'Landbrugsstøtteordninger i EF og USA', *Tidsskrift for landøkonomi*, vol. 173, no. 3, pp. 193-201.

Knoke, David (1990), *Political Networks: The Structural Perspective*, Cambridge University Press: Cambridge.

Knudsen, Tim (1993), *Den danske stat i Europa*, Jurist- og Økonomforbundets Forlag: Copenhagen.

Kommissionen for de europæiske Fællesskaber (1981), *Retningslinier for EF's landbrug: Memorandum, der supplerer Kommissionens rapport vedrørende mandatet af 30. maj 1980*, COM(81) 608, Kontoret for De europæiske Fællesskabers officielle Publikationer: Luxembourg.

Kommissionen for De europæiske Fællesskaber (1985), *Perspektiver for den fælles landbrugspolitik: Kommissionens Grønbog*, COM(85) 333. Kontoret for De europæiske Fællesskabers officielle Publikationer: Luxembourg.

Kommissionen for de europæiske Fællesskaber (1991), *Den fælles landbrugspolitiks udvikling og fremtid - Kommissionens betragtninger*, COM(91) 100, Kontoret for De europæiske Fællesskabers officielle Publikationer: Luxembourg.

Kommissionen for de europæiske Fællesskaber (1991a), *Den fælles landbrugspolitiks udvikling og fremtid: Opfølgning af dokumentet med Kommissionens betragtninger, KOM(91) 100 af 1. februar 1991*, COM(91) 258, Kontoret for De europæiske Fællesskabers officielle Publikationer: Luxembourg.

Krasner, Stephen D. (1984), 'Approaches to the State. Alternative Conceptions and Historical Dynamics', *Comparative Politics*, vol. 16, no. 2, pp. 223-246.

Land, various issues.

Landboforeninger, De danske (1973), *Beretning 1972-73*, Copenhagen.

Landboforeninger, De danske (1974), *Beretning 1973-74*, Copenhagen.

Landboforeninger, De danske (1975), *Beretning 1974-75*, Copenhagen.

Landboforeninger, De danske (1978), *Beretning 1978-79*, Copenhagen.

Landboforeninger, De danske (1979), *Beretning 1978-79*, Copenhagen.

Landboforeninger, De danske (1980), *Beretning 1979-80*, Copenhagen.

Landboforeninger, De danske (1982), *Beretning 1981-82*, Copenhagen

Landboforeninger, De danske (1984), *Beretning 1984*, Copenhagen.

Landboforeninger, De danske (1985), *Beretning 1985*, Copenhagen.

Landboforeninger, De danske (1986), *Beretning 1986*, Copenhagen.

Landboforeninger, De danske (1987), *Beretning 1987*, Copenhagen.

Landboforeninger, De danske (1987a), *Landbrugets aktionsplan for et renere havmiljø. Dokumentation*, Copenhagen.

Landboforeninger, De danske (1987b), 'Vedrørende regeringens handlingsplan mod forurening af det danske vandmiljø med næringsstoffer', Copenhagen, April 2, 1987, published in *Folketingstidende B* 'Bilagshæfte til Beretning over Vandmiljøplanen afgivet af miljø- og planlægningsudvalget den 30. april 1987', Copenhagen, April 2, 1987, pp. 118-25.

Landboforeninger, De danske (1988), *Beretning 1988*, Copenhagen.

Landboforeninger, De danske (1990), *Beretning 1990*, Copenhagen.

Landboforeninger, De danske (1991), *Beretning 1991*, Copenhagen.

Landboforeninger, De danske (1991a), *En bæredygtig udvikling i landbruget*, Copenhagen.

Landbrugets informationskontor (1984), *Gylle - håndtering og anvendelse*, Greve Strand.

Landbrugsministeriet (1977), *Fra forslag til funktion i EFs landbrugspolitik*, Copenhagen.

Landbrugsministeriet (1984), *Kvælstoftilførsel og kvælstofudvaskning i dansk planteproduktion. Gennemsnitsopgørelser for perioden 1978-1982*, Copenhagen.

Landbrugsministeriet (1986), *En miljøvenlig landbrugsproduktion - et strategioplæg*, Copenhagen.

Landbrugsministeriet, (1987), *Landbrugets fremtid: Landbrugsministeriets konference 26. februar 1987*, Copenhagen.

Landbrugsministeriet (1991), *Bæredygtigt landbrug: En teknisk redegørelse*, Copenhagen.

Landbrugsministeriet (1991a), 'Referat af 879. møde i Specialkomiteen for Landbrug, den 4-5 november 1991', Copenhagen.

Landbrugsministeriet (1991b), 'Referat af 880. møde i Specialkomiteen for Landbrug, den 11-12 november 1991', Copenhagen.

Landbrugsministeriet (1991c), 'EF Repræsentationens referat af rådsmøde

(Landbrug) 18-19 november 1991', Copenhagen.

Landbrugsministeriet (1991d), 'Referat af 881. møde i Specialkomiteen for Landbrug, den 25-26 november 1991', Copenhagen.

Landbrugsministeriet (1991e), 'EF Repræsentationens referat af rådsmøde (Landbrug) 11-12 december 1991', Copenhagen.

Landbrugsministeriet (1992), 'EF-Repræsentationens referat af rådsmøde (Landbrug) den 27.-29. januar 1992', Copenhagen.

Landbrugsministeriet (1992a), 'EF-Repræsentationens referat af rådsmøde (Landbrug) den 10.-11. februar 1992', Copenhagen.

Landbrugsministeriet (1992b), 'EF-Repræsentationens referat af rådsmøde (Landbrug) den 2-3 marts 1992', Copenhagen.

Landbrugsministeriet (1992c), 'EF-Repræsentationens referat af rådsmøde (Landbrug) den 20 marts 1992', Copenhagen.

Landbrugsministeriet (1992d), 'EF-Repræsentationens referat af rådsmøde (Landbrug) den 30-31 marts 1992', Copenhagen.

Landbrugsministeriet (1992e), 'EF-Repræsentationens referat af rådsmøde (Landbrug) den 28-29 april 1992', Copenhagen.

Landbrugsministeriet (1992f), 'Referat af 894. møde i Specialkomiteen for Landbrug den 11-12 maj 1992', Copenhagen.

Landbrugsministeriet (1992g), 'EF-Repræsentationens referat af rådsmøde (Landbrug) den 18-21 maj 1992', Copenhagen.

Landbrugsraadet and Miljøministeriet (1980), *Landbrug og miljø. Konferencegrundlag og indlæg på landbrugsorganisationernes og miljøministeriets konference i Rønne, september 1980*, Landbrugsraadet: Copenhagen.

Landsbladet, 9 October 1992, 15 March 1996.

Lane, Jan-Erik and Ersson, Svante O. (1991), *Politics and Society in Western Europe*, (2nd ed.), Sage: London.

Lantbruksstyrelsen (1990), *Lagring och spridning av stallgödsel - Ytterligare restriktioner till skydd för miljön*, Lantbruksstyrelsens rapport 1990: 2, Jönköping.

Lantbruksstyrelsen (1990a), *Grön mark - Utvidgad höst- och vinterbevuxen mark*, Lantbruksstyrelsens rapport 1990: 3, Jönköping.

Lantbruksstyrelsen (1990b), *Förslag till typgodkännande av gödselspridare - Utredning TYP 90*, Lantbruksstyrelsens rapport 1990: 4, Jönköping.

Larsen, Arne (1992), 'GATT og/eller MacSharry', *Jord og viden*, no. 7, pp. 6-7.

Larsen, Helge (ed.) (1980), *Det Radikale Venstre i medvind og modvind 1955-1980*, Tidens Tankers Forlag: Copenhagen.

Larsson, Hans Albin (1980), *Partireformation - från bondeförbund till centerparti*, CWK Gleerup: Lund.

Larsson, Torbjörn (1986), *Regeringen och dess kansli: Samordning och byråkrati i maktens centrum*, Studentlitteratur: Lund.

Larsson, Torbjörn (1993), *Det svenska statsskicket*, Studentlitteratur: Lund.

Laumann, Edward O., Knoke, David and Kim, Yong-Hak (1985), 'An Organizational Approach to State Policy Formation: A Comparative Study of Energy and Health Domains', *American Sociological Review*, vol. 50, no. 1, pp. 1-19.

Laumann, Edward O. and Knoke, David (1987), *The Organizational State: Social Choice in National Policy Domains*, The University of Wisconsin Press: Madison.

Le Heron, Richard (1993), *Globalized Agriculture: Political Choice*, Pergamon Press: Oxford.

Lembruch, Gerhard (1979), 'Liberal Corporatism and Party Government' in Philippe C. Schmitter and Gerhard Lembruch (eds), *Trends Toward Corporatist Intermediation*, Sage: London, pp. 147-83.

Lembruch, Gerhard (1991), 'The Organization of Society, Administrative Strategies, and Policy Networks', in Roland Czada and Adrienne Windhoff-Héritier (eds), *Political Choice: Institutions, Rules, and the Limits of Rationality*, Campus Verlag/Westview Press: Frankfurt am Main/Boulder, pp. 121-58.

Leone, Robert A. (1986), *Who Profits. Winners, Losers and Government Regulation*, Basic Books: New York.

Lewin, Leif (1984), *Ideologi och strategi: Svensk politik under 100 år*, P.A. Nordstedts & Söners förlag: Stockholm.

Lindblad, Ingemar, Stålvant, Carl-Einar, Wahlbäck, Krister, Wiklund, Claes (1984), *Politik i Norden: En jämförande översikt*, Liber Förlag: Stockholm.

Linder, Stephen, H. and Peters, B. Guy (1989), 'Instruments of Government: Perceptions and Contexts', *Journal of Public Policy*, vol. 9, no. 1, pp. 35-58.

Lindrup, Kurt (1986), *Implementering og administration af EFs Mælkekvoteordning i Danmark*, Department of Political Science, Aarhus University, (unpublished MA dissertation).

Lodge, Juliet (1989), 'EC Policymaking: Institutional Considerations', in Juliet Lodge (ed.), *The European Community and the Challenge of the Future*, Pinter Publishers: London.

Lov no. 595, 1972, 'Lov om administration af Det europæiske Fællesskabs forordninger om markedsordninger for landbrugsvarer m.v.', in *Lovtidende A*, Copenhagen.

Lov no. 372, 1973, 'Lov om miljøbeskyttelse', in *Lovtidende A*, Copenhagen.

Lov no. 16, 1987, 'Lov om støtte til miljøforbedrende investeringer i mindre landbrug m.v.', in *Lovtidende A*, Copenhagen.

Lov no. 1172, 1992, 'Bekendtgørelse af lov om støtte til miljøforbedrende investeringer i mindre landbrug m.v.', in *Lovtidende A*, Copenhagen.

Lowi, Theodore J. (1964), 'American Business, Public Policy, Case Studies, and Political Theory', *World Politics*, vol. 16, no. 4, pp. 677-715.

LRF (1972), *Information special: Livmedlen och miljön*, Stockholm.

LRF (1985), 'Prisa kvalitet: Program för kvalitetsproduktion av svenska livsmedel råvaror - Problembeskrivning och åtgärder', 9/85/forskn/kp, Stockholm.

LRF (1990), *Den värld vill vi ha: Bonden och miljön*, LTs förlag: Stockholm.

LRF (1990a), 'Yttrande över rapport ang. En ny livsmedelpolitik (Ds 1989: 63)', Dnr 1170/89, 1990-02-19, Stockholm.

LRF (1990b), 'En framtidsinriktad näringspolitik', bilaga 1 till 'Yttrande över rapport ang. En ny livsmedelpolitik (Ds 1989: 63)', Dnr 1170/89, 1990-02-19,

Stockholm.

LRF (1992), *Sveriges bönder - steget före: Konsumentpolitiskt handlingsprogram*, Stockholm.

Lukes, Steven (1974), *Power: A Radical View*, Macmillan: London.

Lundqvist, Lennart J. (1971), *Miljövårdsförvaltning och politisk struktur*, Prisma: Lund.

Lundqvist, Lennart J. (1996), 'Sweden', in Peter Munk Christiansen (ed.), *Governing the Environment: Politics, Policy and Organization in the Nordic Countries*, Nord 1996:5, Nordic Council of Ministers, Copenhagen, pp. 257-336.

Madsen, Jens Astrup (1987), *Landbrug og miljø*, Department of Political Science, Aarhus University, (unpublished MA dissertation).

Mahler, Vincent A. (1991), 'Domestic & International Sources of Trade Policy: The Case of Agriculture in the European Community & the United States', *Polity*, vol. 24, no. 1, pp. 27-47.

Majone, Giandomenico (1989), *The Evidence, Argument & Persuasion in the Policy Process*, Yale University Press: New Haven.

Maloney, William A. and Richardson, Jeremy J. (1994), 'Water Policy Making in England and Wales: Policy Communities Under Pressure?', *Environmental Politics*, vol. 3, no. 4, pp. 110-38.

Maloney, William A., Jordan, Grant and McLaughlin, Andrew. M (1994), 'Interest Groups and Public Policy: The Insider/Outsider Model Revisited', *Journal of Public Policy*, vol. 14, no. 1, pp. 17-38.

Manegold, Dirk (1991), 'EC Agricultural Policy in 1990/91: Growing Need for Real Policy Reform' *Review of Marketing and Agricultural Economics*, vol. 59, no. 2, pp. 98-129.

Manegold, Dirk (1993), 'EC Agricultural Policy in 1992-93: Implementation of CAP Reform', *Review of Marketing and Agricultural Economics*, vol. 61, no. 2, pp. 113-40.

March, James G. and Olsen, Johan P. (1989), *Rediscovering Institutions: The Organizational Basis of Politics*, The Free Press: New York.

Marin, Bernd and Mayntz, Renate (1991), 'Introduction: Studying Policy Networks', in Bernd Marin and Renate Mayntz (eds), *Policy Networks: Empirical Evidence and Theoretical Considerations*, Campus Verlag/Westview Press: Frankfurt am Main/Boulder, pp. 11-23.

Marsh, David (1992), 'Youth Employment Policy 1970-1990: Towards the Exclusion of Trade Unions', in David Marsh and R.A.W. Rhodes (eds), *Policy Networks in British Government*, Clarendon Press: Oxford, pp. 167-99.

Marsh, David (1994), 'Coming of Age, but not Learning from Experience: One Cheer for Rational Choice Theory', Department of Government, University of Strathclyde.

Marsh, David (1995), 'State Theory and the Policy Network Model', Department of Government, University of Strathclyde.

Marsh, David and Daugbjerg, Carsten (1997), 'Linking the Policy Network Model with Micro and Macro Level Analysis', in David Marsh (ed.), *Policy Networks:*

Theory and Comparison, Open University Press: Milton Keynes (forthcoming).

Marsh, D. and Locksley, G. (1983), 'Capital: The Neglected Face of Power?', in David Marsh (ed.), *Pressure Politics: Interest Groups in Britain*, Junction Books: London

Marsh, David and Rhodes, R.A.W. (1992), 'Policy Communities and Issue Networks: Beyond Typology', in David Marsh and R.A.W. Rhodes (eds), *Policy Networks in British Government*, Clarendon Press: Oxford, pp. 249-68.

Marsh, David and Rhodes, R.A.W. (eds) (1992a), *Policy Networks in British Government*, Clarendon Press: Oxford.

Marsh, David, and Smith, Martin J. (1995), 'The Role of Networks in an Understanding of Whitehall: Towards a Dialectical Approach' Paper presented at Political Studies Association, Annual Conference, University of York, April 18-20.

Marsh, David and Smith, Martin J. (1996), 'Understanding Policy Networks: Towards a Dialectical Approach', Department of Political Science and International Studies, University of Birmingham, (unpublished manuscript).

Mazey, Sonia and Richardson, Jeremy J. (1992), 'British Pressure Groups in the European Community: The Challenge of Brussels', *Parliamentary Affairs*, vol. 45, no. 1, pp. 92-107.

Mazey, Sonia and Richardson, Jeremy J. (1992a), 'Environmental Groups and the EC: Challenges and Oppurtunities', *Environmental Politics*, vol. 1, no. 4, pp. 109-28.

Mazey, Sonia and Richardson, Jeremy J. (1993), 'Interest Groups in the European Community', in Jeremy J. Richardson (ed.), *Pressure Groups*: Oxford University Press: Oxford, pp. 191-213.

Mazey, Sonia and Richardson, Jeremy J. (1993a), 'Introduction: Transference of Power, Decision Rules and Rules of the Game', in Sonia Mazey and Jeremy J. Richardson (eds), *Lobbying in the European Community*, Oxford University Press: Oxford, pp. 3-26.

Mazey, Sonia and Richardson, Jeremy J. (1993b), 'Conclusion: A European Policy Style', in Sonia Mazey and Jeremy J. Richardson (eds), *Lobbying in the European Community*, Oxford University Press: Oxford, pp. 246-58.

McFarland, Andrew S. (1992), 'Interest Groups and the Policymaking Process: Sources of Countervailing Power in America', in Mark P. Petracca (ed.), *The Politics of Interest: Interest Groups Transformed*, Westview Press: Boulder, pp. 58-79.

Meester, Gerrit and van der Zee, Frans A. (1993), 'EC Decision-making, Institutions and the Common Agricultural Policy', *European Review of Agricultural Economics*, vol. 20, no. 2, pp. 131-50.

Micheletti, Michele (1990), *The Swedish Farmers' Movement and Government Agricultural Policy*, Praeger: New York.

Miljø- og planlægningsudvalget (1987), 'Beretning om Vandmiljøplanen', in *Folketingstidende*, tillæg B, Copenhagen.

Miljøministeriet (1980), *Landbrug og miljø: Miljøministeriets udredningsinsats og forskningsbehov*, Copenhagen.

Miljøministeriet (1986), 'Aktionsplan for havet omkring Danmark', in *Folketingstidende*, tillæg B 'Bilagshæfte til Beretning over Vandmiljøplanen afgivet af miljø- og planlægningsudvalget den 30. april 1987', pp. 5-8.

Miljøministeriet (1987), 'Handlingsplan mod forurening af det danske vandmiljø med næringssalte', in *Folketingstidende*, tillæg B 'Bilagshæfte til Beretning over Vandmiljøplanen afgivet af miljø- og planlægningsudvalget den 30. april 1987', pp. 9-21.

Miljøstyrelsen (1984), *NPO-redegørelsen*, Copenhagen.

Miljøstyrelsen (1985), *Miljøafgifter: Mulighederne for at anvende afgifter som supplerende styringsmiddel/finansiering af nye foranstaltninger på miljøområdet*, Orientering fra Miljøstyrelsen, no. 5 1985, Copenhagen.

Miljøstyrelsen (1986), 'Rapport fra ekspertgruppe vedrørende reduktion af landbrugets kvælstofudledning' (including separate statements), M 86-11-122, 16 December, 1986, Copenhagen.

Miljøstyrelsen (1992), *Vandmiljø-92*, redegørelse fra Miljøstyrelsen nr. 2 1992, Copenhagen.

Mills, Michael (1992), 'The Case of Food and Health and the Use of Network Analysis', in David Marsh and R.A.W. Rhodes (eds), *Policy Networks in British Government*, Clarendon Press: Oxford, pp. 149-66.

Mills, Mike and Saward, Michael (1994), 'All Very Well in Practice, But What About the Theory? A Critique of the British Idea of Policy Networks', in Patrick Dunleavy and Jeffrey Stanyer (eds), *Contemporary Political Studies*, Political Studies Association: Belfast, pp. 79-92.

Mitnick, Barry M. (1993), 'The Strategic Uses of Regulation - And Deregulation', in Barry M. Mitnick (ed.), *Corporate Political Agency: The Construction of Competition in Public Affairs*, Sage: London, pp. 67-89.

Moyer, H. Wayne and Josling, Timothy E. (1990), *Agricultural Policy Reform: Politics and Processes in the EC and the USA*, Harvester Wheatsheaf: London.

Murphy, Jerome T. (1980), *Getting the Facts: a Fieldwork Guide for Evaluators and Policy Analysts*, Scott, Foresmann and Company: Glenview, Illinois.

National Consumer Council (1989), 'Consumers and the Common Agricultural Policy', *Journal of Consumer Policy*, vol. 12, no. 2, pp. 165-92.

National Consumer Council (1995), *Agricultural Policy in the European Union: The Consumer Agenda for Reform*, London.

Naturvårdsverket (1970), *Riktlinier för miljöskyddande åtgärder vid animalieproduktion*, Naturvårdsverkets publikationer 1970: 3, Solna.

Naturvårdsverket (1987), *Aktionsplan mot havsföroreningar*, Solna.

Naturvårdsverket (1990), *Hav '90: Aktionsprogram mot havsföroreningar*, Solna.

Nedergaard, Peter (1988), *EF's landbrugspolitik under omstilling*, Jurist- og økonomforbundets forlag: Copenhagen.

Nedergaard, Peter (1994), 'The Political Economy of CAP Reform', in Rasmus Kjeldahl and Michael Tracy (eds), *Renationalisation of the Common Agricultural Policy?* Institute of Agricultural Economics/Agricultural Policy Studies: Valby/la Hutte, pp. 85-103.

Nedergaard, Peter (1995), 'The Political Economy of CAP Reform', in Finn Laursen (ed.), *The Political Economy of European Integration*, Kluwer Law International: The Hague, pp. 111-44.

Nedergaard, Peter, Hansen, Henning Otte og Mikkelsen, Preben (1993), *EF's landbrugspolitik og Danmark: Udvikling frem til år 2000*, Handelshøjskolens Forlag: Copenhagen.

Nello, Susan Senior (1989), 'European Interest Groups and the CAP', *Food Policy*, vol. 14, no. 2, pp. 101-6.

Neville-Rolfe, Edmund (1984), *The Politics of Agriculture in the European Community*, European Centre for Political Studies: London.

Nordic Council of Ministers (1992), *Leaching and Runoff of Nitrogen from Agricultural Areas in the Nordic Countries*, Report no. 1992:594, Nordic Council of Ministers: Copenhagen.

Nordic Council of Ministers (1992a), *Leaching and Runoff of Nitrogen from Agricultural Areas in the Nordic Countries: Appendix 2: Workshops on N-losses from Agricultural Areas*, Report no. 1992:595, Nordic Council of Ministers: Copenhagen.

Nordisk Ministerråd (1992), *Landbruk og miljø i Norden - arbejdsplan 1992-1995*, Nordiske seminar- og arbejdsrapporter 1992: 522, Nordisk Ministerråd: Copenhagen.

Nordisk Ministerråd (1993), *Mod strømmen... - en vurdering af styringsmidler overfor afstrømningen af næringssalte fra landbruget i de nordiske lande*, Nordiske seminar- og arbejdsrapporter 1993: 565, Nordisk Ministerråd: Copenhagen.

Nordlinger, Eric A. (1981), *On the Autonomy of the Democratic State*, Harvard University Press: Cambridge, Mass.

Nordlinger, Eric A. (1988), 'The Return to the State: Critiques', *American Political Science Review*, vol. 82, no. 3, pp. 875-85.

Norton, Philip (1991), 'The Changing Face of Parliament: Lobbying and its Consequences', in Philip Norton (ed.), *New Directions in British Politics? Essays on the Evolving Constitution*, Edward Elgar: Aldershot, pp. 58-82.

Norton, Philip (1993), *Does Parliament Matter*, Harvester Wheatsheaf: London.

Nugent, Niell (1991), *The Government and Politics of the European Community* (2nd ed.), Macmillan: London.

Nugent, Neill (1994), *The Government and Politics of the European Union* (3rd ed.), Macmillan: London.

Nørgaard, Asbjørn Sonne (1994), 'Institutions and Post-Modernity in IR: The "New" EC', *Cooperation and Conflict*, vol. 29, no. 3, pp. 245-87.

Nørgaard, Asbjørn Sonne (1996), 'Rediscovering Reasonable Rationality in Institutional Analysis', *European Journal of Political Research*, vol. 29, January, pp. 31-57.

OECD (1989), *Agricultural and Environmental Policies: Opportunities for Integration*, OECD: Paris.

OECD (1989a), *Economic Instruments for Environmental Protection*, OECD: Paris.

OECD (1993), *Agricultural and Environmental Policy Integration: Recent Progress and New Directions*, OECD: Paris.

OECD (1994), *Agricultural Policies, Markets and Trade: Monitoring and Outlook 1994*, OECD: Paris.

OECD (1995), *Agricultural Policy Reform and Adjustment: The Swedish Experience*, OECD: Paris.

Official Journal L 26, 03/02/92, 92/41/EEC, Euratom, ECSC: Final adoption of the general budget of the European Communities for the financial year 1992, Brussel.

Official Journal L 181 01/07/92, 'Council Regulation (EEC) no. 1766 of 30 June 1992 on the common organization of the market in cereals', Brussels.

Olsen, Johan P. (1983), *Organized Democracy: Political Institutions in a Welfare State: The Case of Norway*, Universitetsforlaget: Bergen.

Olsen, Johan P. (1991), 'Political Science and Organization Theory: Parallel Agendas but Mutual Disregard', in Roland Czada and Adrienne Windhoff-Héritier (eds), *Political Choice: Institutions, Rules, and the Limits of Rationality*, Campus Verlag/Westview Press: Frankfurt am Main/Boulder, pp. 87-119.

Olsen, Johan P., Roness, Paul and Sætren, Harald (1982), 'Norway: Still Peaceful Coexistence and Revolution in Slow Motion?', in Jeremy Richardson (ed.), *Policy Styles in Western Europe*, George Allan & Unwin: London.

Olson, Mancur (1965), *The Logic of Collective Action*, Cambridge University Press: Cambridge.

Ordbruk no. 7, August 1994.

Ostrom, Elinor (1990), *Governing the Commons: The Evolution of Institutions for Collective Action*, Cambridge University Press: Cambridge.

Ostrom, Elinor (1991), 'Rational Choice Theory and Institutional Analysis: Towards Complementarity', *American Political Science Review*, vol. 85, no. 1, pp. 237-43.

Panebianco, Angelo (1988), *Political Parties: Organisation and Power*, Cambridge University Press: Cambridge.

Pedersen, Mogens N. (1968), 'Rekruttering af danske folketingsmænd', paper presented for the 2nd Nordic Political Science Conference in Helsinki, 19-21 August 1968.

Pedersen, Mogens N. (1987), 'The Danish "Working Multiparty System": Break Down or Adaption', in Hans Daalder (ed.), *Party Systems in Denmark, Austria, Switzerland, the Netherlands, and Belgium*, Frances Pinter: London, pp. 1-60.

Pedersen, Uffe Toudal (1981), *Landbrugsministeriets udvikling og samspil med organisationerne og ministeriet efter 1973*, Department of Political Science, Aarhus University (unpublished MA dissertation).

Pedersen, Jørgen Flindt and Geckler, Rolf (1987), 'Græsrødder og embedsmænd', in Rolf Geckler and Jørgen Grønnegaard Christensen (eds), *På ministerens vegne: På vej mod en embedsmandsstat?*, Gyldendal: Copenhagen.

Peters, B. Guy (1994), 'Agenda-setting in the European Community', *Journal of European Policy*, vol. 1, no. 1, June, pp. 9-26.

Peterson, John (1992), 'The European Technology Community: Policy Networks in a Supranational Setting', in David Marsh and R.A.W. Rhodes (eds), *Policy Networks in British Government*, Clarendon Press: Oxford, pp. 226-48.

Peterson, John (1995), 'Decision-making in the European Union: Towards a Framework for Analysis', *Journal of European Public Policy*, vol. 2, no. 1, pp. 69-93.

Peterson, John (1995a), 'Policy Networks and European Union Policy Making: A Reply to Kassim', *West European Politics*, vol. 18, no. 2, pp. 389-407.

Petersson, Olof (1989), *Maktens nätverk: En undersökning av regeringskansliets kontakter*, Carlsons Bokförlag: Stockholm.

Petersson, Olof and Söderlind, Donald (1993), *Forvaltningspolitik* (2nd ed.), Publica: Stockholm.

Phillips, Peter W.B. (1990), *Wheat, Europe and the GATT: A Political Economy Analysis*, Pinter Publishers: London.

Press, no. 37, December 1988.

Proposition 1969: 28, 'Kungl. Maj:ts proposition til Riksdagen med förslag till milöskyddslag m.m.', Stockholm.

Proposition 1972: 79, 'Kungl. Maj:ts proposition angående regleringen av priserna på jordbruksprodukter m.m.', Stockholm.

Proposition 1983/84: 176, 'Regeringens proposition 1983/84: 176 om avgifter på gödsels- och bekämpningsmedel', Stockholm.

Proposition 1984/85: 10, 'Regeringens proposition 1984/85: 10 om ändring i miljöskyddslagen (1969: 387) m.m.', Stockholm.

Proposition 1984/85: 166, 'Regeringens proposition 1984/85: 166, om livsmedelpolitiken'.

Proposition 1986/87: 95, 'Regeringens proposition 1986/87: 95. Totalförsvarets fortsatta utveckling', Stockholm.

Proposition 1986/87: 146, 'Regeringens proposition 1986/87: 146 om reglering av priserna på jordbruksprodukter, m.m.', Stockholm.

Proposition 1987/88: 85, 'Regeringens proposition 1987/88: 85 om miljöpolitikken inför 1990-talet', Stockholm.

Proposition 1987/88: 128, 'Regeringens proposition 1987/88: 128 om miljöförbättrande åtgärder i jordbruket, m.m.', Stockholm.

Proposition 1988/89: 47, 'Regeringens proposition 1988/89: 47 om vissa ekonomisk-politiska åtgärder, m.m.', Stockholm.

Proposition 1988/89: 154, 'Regeringens proposition 1988/89: 154 om en ny regional statslig förvaltning', Stockholm.

Proposition 1989/90: 146, 'Regeringens proposition 1989/90: 146 om livsmedelpolitiken', Stockholm.

Proposition 1989/90: 146, 'Regeringens proposition 1989/90: 146 om livsmedelpolitiken. Bilagedel', Stockholm.

Proposition 1990/91: 90, 'Regeringens proposition 1990/91: 90. En god livsmiljö', Stockholm.

Raadsnyt no. 22, 21 June 1991.

Rabinowicz, Ewa (1993), 'Swedish Agricultural Policy Reform: Are There Lessons to be Learned for the CAP?', in *EC Agricultural Policy by the End of the Cen-*

tury: *Proceedings of the 28ᵗʰ Seminar of the European Association of Agricultural Economists, September 10-12, 1992 in Lisbon*, Wissenschaftsverlag Vauk: Kiel, pp. 275-93.

Ragin, Charles (1987), *The Comparative Method: Moving Beyond Qualitative and Quantitative Strategies*, California University Press: Berkeley.

Rasmussen, Erik (1972), *Komparativ Politik 2*, Gyldendal: Copenhagen.

Rayner, A.J, Ingersent K.A. and Hine, R.C. (1993), 'Agricultural Trade and the GATT', in A.J. Rayner and David Coleman (eds) *Current Issues in Agricultural Economics*, Macmillan: London, pp. 62-95.

Read, Melvin D. (1992), 'Policy Networks and Issue Networks: The Politics of Smoking', in David Marsh and R.A.W. Rhodes (eds), *Policy Networks in British Government*, Clarendon Press: Oxford, pp. 124-48.

Rehling, David (1994), 'Miljøorganisationer mellem medlemmer og velfærdssamfund', in Johannes Michelsen (ed.), *Medlemsbaserede organisationer i velfærdssamfundet - nutid og fremtid*, Årsskrift 1994, Foreningen for Andelsstudier: Albertslund, pp. 83-95.

Rhodes, R.A.W. (1981), *Control and Power in Central-Local Government Relations*, Gower Publishing: Farnborough.

Rhodes, R.A.W. (1986), *The National World of Local Government*, Allen & Unwin: London.

Rhodes, R.A.W. (1990), 'Policy Networks: A British Perspective', *Journal of Theoretical Politics*, vol. 2, no. 3, pp. 293-317.

Rhodes, R.A.W. (1992), *Beyond Westminster and Whitehall: The Sub-central Governments of Britain*, Routledge: London, [first published in 1988 by Unwin Hyman, London].

Rhodes, R.A.W. and Marsh, David (1992), 'Policy Networks in British Politics: A Critique of Existing Approaches', in David Marsh and R.A.W. Rhodes (eds), *Policy Networks in British Government*, Clarendon Press: Oxford, pp. 1-25.

Rhodes, R.A.W. and Marsh, David (1992a), 'New Directions in the Study of Policy Networks', *European Journal of Political Research*, vol. 21, no 1-2, pp. 181-205.

Rhodes, R.A.W. and Marsh, David (1994), 'Policy Networks: "Defensive" Comments, Modest Claims, and Plausible Research Strategies', paper to the PSA Annual Conference, University of Swansea, 29-31 March 1994.

Richardson, Jeremy and Jordan, A. G. (1979), *Governing under Pressure: The Policy Process in a Post-Parliamentary Democracy*, Martin Robertson: Oxford.

Richardson, Jeremy J., Maloney, William A. and Rüdig, Wolfgang (1992), 'The Dynamics of Policy Change: Lobbying and Water Privatization', *Public Administration*, vol. 70, pp. 157-75.

Rieger, Elmer (1996), 'The Common Agricultural Policy', in Helen Wallace and William Wallace (eds) *Policy-Making in the European Union* (3ʳᵈ ed.), Oxford University Press: Oxford, pp. 97-123.

Riksdagen (1990), 'Votering 1132, 1990-06-09', Stockholm.

Ritson, Christopher (1991), 'The CAP and the Consumer', in Christopher Ritson and

David Harvey (eds), *The Common Agricultural Policy and the World Economy: Essays in Honour of John Ashton*, CAB International: Wallingford, pp. 119-37.

Robinson, Alan David (1961), *Dutch Organised Agriculture in International Politics*, Martinus Nijhoff/'S-Gravenhage: The Hague.

Roche, M.M., Johnston, T. and Le Heron, R.B. (1992), 'Farmers' interest groups and agricultural policy in New Zealand during the 1980s', *Environment and Planning A*, vol. 24, no. 12, pp. 1749-67.

Rommetvedt, Hilmar (1992), 'Norway: From Consensual Majority Parliamentarism to Dissensual Minority Parliamentarism', in Erik Damgaard (ed.), *Parliamentary Change in the Nordic Countries*, Scandinavian University Press: Oslo, pp. 51-97.

Rose, Richard (1991), 'Comparing Forms of Comparative Analysis', *Political Studies*, vol. 39, no. 3, pp. 446-62.

Rothstein, Bo (1988), 'Aktör-Strukturansatsen: Ett metodiskt dilemma', *Statsvetenskaplig Tidsskrift*, vol. 91, pp. 27-40.

Rothstein, Bo (1992), *Den korporativa Staten: Intresseorganisationer och statsförvaltning i svensk politik*, Norstedts: Stockholm.

Sabatier, Paul A. (1986), 'Top-Down and Bottum-Up Approaches to Implementation Research: a Critical Analysis and Suggested Synthesis', *Journal of Public Policy*, vol. 6, no. 1, pp. 21-48.

Sabatier, Paul A. (1993), 'Policy Change over a Decade or More', in Paul A. Sabatier and Hank C. Jenkins-Smith (eds), *Policy Change and Learning: An Advocacy Coalition Framework*, Westview Press: Boulder, pp. 13-39.

Sabatier, Paul A. and Mazmanien, Daniel A. (1981), 'The Implementation of Public Policy: A Framework for Analysis', in Paul A. Sabatier and Daniel A. Mazmanien (eds), *Effective Policy Implementation*, Lexington Books/D.C. Heath & co: Lexington, Massachusetts, pp. 3-35.

Salamon, Lester M. (1981), 'Rethinking Public Management: Third-Party Government and the Changing Forms of Government Action' *Public Policy*, vol. 29, no. 3, pp. 255-75.

Salamon, Lester M. and Michael S. Lund (1989), 'The Tools Approach: Basic Analytics', in Lester M. Salamon (ed.), *Beyond Privatization: The Tools of Government Action*, The Urban Institute Press: New York, pp. 23-49.

Sandrey, Ron and Reynolds, Russell (eds) (1990), *Farming without Subsidies: New Zealand's Recent Experience*, MAF and GP Books: New Zealand.

Sannerstedt, Anders and Sjölin, Mats (1992), 'Sweden: Changing Party Relations in a More Active Parliament', in Erik Damgaard (ed.) *Parliamentary Change in the Nordic Countries*, Scandinavian University Press: Oslo, pp. 99-149.

Sartori, Giovanni (1976), *Parties and Party Systems: A framework for Analysis*, Volume I, Cambridge University Press: Cambridge.

Saward, Michael (1992), 'The Civil Nuclear Network in Britain', in David Marsh and R.A.W. Rhodes (eds), *Policy Networks in British Government*, Clarendon Press: Oxford, pp. 75-99.

Schattschneider, E.E. (1960), *The Semisovereign People: A Realist's View of Democracy in America*, The Dryden Press: Hinsdale, Ill.

Schlozman, K.L. and Tierney, J.T. (1986), *Organized Interests and American Democracy*, Harper and Row Publishers: New York.

Schmitter, Phillipe C. (1974), 'Still the Century of Corporatism?', *Review of Politics*, vol. 36, pp. 85-131.

Schmitter, Phillipe C. (1979), 'Modes of Interest Intermediation and Models of Societal Change in Western Europe', in Philippe C. Schmitter and Gerhard Lembruch (eds), *Trends Towards Corporatist Intermediation*, Sage: London pp. 63-94.

Schneider, Volker (1992), 'The Structure of Policy Networks: A Comparison of the "Chemicals Control" and "Telecommunications" Policy Domain in Germany', *European Journal of Political Research*, vol. 21, no. 1-2, pp. 109-29.

Seyler, Hans (1983), *Hur bonden blev lönarbetare: Industrisamhället och den svenska bondeklassens omvandling*, Arkiv: Lund.

SFS 1969: 387, 'Miljöskyddslag', Stockholm.

SFS 1969: 388, 'Kungl. Maj:ts Miljöskyddskungörelse', Stockholm.

SFS 1972: 224, 'Kungl. Maj:ts kungörelse om ändring i miljöskyddskungörelse', Stockholm.

SFS 1972: 293, 'Kungl. Maj:ts kungörelse om statsbidrag till miljövårdande åtgärder inom jordbruks- och trädgårdforetag', Stockholm.

SFS 1988: 640, 'Förordning om ändring i förordningen om skötsel av jordbruksmark', Stockholm.

SFS 1989: 12, 'Förordning om statsbidrag till miljöforbättrande åtgärder i jordbruket', Stockholm.

Skocpol, Theda (1985), 'Bringing the State Back In: Strategies of Analysis in Current Research', in Peter Evans, Dietrich Rueschemeyer and Theda Skocpol (eds), *Bringing the State Back In*, Cambridge University Press: Cambridge, pp. 3-37.

Skocpol, Theda and Finegold, Kenneth (1982), 'State Capacity and Economic Intervention in the Early New Deal', *Political Science Quarterly*, vol. 97, no. 2, pp. 255-78.

Smith, Martin J. (1990), 'Pluralism, Reformed Pluralism and Neopluralism: the Role of Pressure Groups in Policy-Making', *Political Studies*, vol. 38, no. 2, pp. 302-22.

Smith, Martin J. (1990a), *The Politics of Agricultural Support in Britain: The Development of the Agricultural Policy Community*, Dartmouth: Aldershot.

Smith, Martin J. (1992), 'The Agricultural Policy Community: Maintaining a Close Relationship', in David Marsh and R.A.W. Rhodes (eds), *Policy Networks in British Government*, Clarendon Press: Oxford, pp. 27-50.

Smith, Martin J. (1993), *Pressure, Power and Policy: State Autonomy and Policy Networks in Britain and the United States*, Harvester Wheatsheaf: London.

SOU 1974: 35, *Spridning av kemiska medel: Betänkande avgivet av udredningen om spridning av kemiska medel*, Allmänna Förlaget: Stockholm.

SOU 1977: 17, *Översyn av jordbrukspolitiken: Betänkande av 1972 års jordbruksutredning*, LiberFölag/Allmänna Förlaget: Stockholm.

SOU 1983: 10, *Använding av växtnäring: Betänkande av udredningen om använ-*

ding av kemiska medel i jord- och skogsbruket, Liber/Allmänna Förlaget: Stockholm.

SOU 1984: 86, *Jordbruks och livsmedelspolitik: Huvudbetänkande av 1983 års livsmedelkommitté*, Liber/Allmänna Förlaget: Stockholm.

SOU 1987: 32, *För en bättre miljö: Betänkande av utredningen om miljövårdens organisation*, Liber/Allmänna Förlaget: Stockholm.

SOU 1987: 44, *Livsmedelpriser och livsmedelkvalitet: Betänkande av 1986 års livsmedelutredning, LMU*, Liber/Allmänna Förlaget: Stockholm.

SOU 1990: 87, *Den nya centrala jordbruksmyndigheten: Slutbetänkande av utredningen om översyn av den centrala myndighetsorganisation på jordbrukets område m.m.*, Allmänna Förlaget: Stockholm.

SOU 1992: 99, *Rådgivning inom jordbruket och trädgårdsnäringen: Betänkande av Rådgivningsutredningen*, Allmänna Förlaget: Stockholm.

Statistisk Sentralbyrå (1993), *Statistisk årbok, 1993*, Oslo.

Statskalender, Sveriges (1983), Liber/Almänna Förlaget: Stockholm.

Steen, Anton (1985), 'The Farmers, the State and the Social Democrats', *Scandinavian Political Studies*, vol. 8, no. 1-2, pp. 45-63.

Steen, Anton (1988), *Landbruket, staten og sosialdemokratene: En komparativ studie av interessekonfliktene i landbrukspolitikken i Norge, Sverige og England 1945-1985*, Universitetsforlaget: Oslo.

Steinberger, Peter J. (1980), 'Typologies of Public Policy: Meaning Construction and the Policy Process', *Social Science Quarterly*, vol. 61, no. 2, pp. 185-97.

Stones, Rob (1992), 'Labour and International Finance, 1964-1967', in David Marsh and R.A.W. Rhodes (eds), *Policy Networks in British Government*, Clarendon Press: Oxford, pp. 200-25.

Svendsen, Gert Tinggaard (1996), *Public Choice and Environmental Regulation: Tradable Permit Systems in the United States and CO_2 Taxation in Europe*, The Aarhus School of Business: Aarhus, (forthcoming on Edward Elgar Publishing).

Svold, Claus (1989), *Danmarks Naturfredningsforening: Fra pæn forening til aggressiv miljøorganisation*, PLS Consult: Aarhus.

Swedish Board of Agriculture (1994), *Swedish Agriculture*, Jönköping.

Swinbank, Alan (1989), 'The Common Agricultural Policy and the Politics of European Decision Making', *Journal of Common Market Studies*, vol. 27, no. 4, June, pp. 303-22.

Swinbank, Alan (1993), 'CAP Reform, 1992', *Journal of Common Market Studies*, vol. 31, no. 3, September, pp. 359-72.

Søborg, Henrik (1983), *Socialdemokratiet og staten: Socialdemokratiets økonomiske politik 1945-72*, SFAH no. 15, Selskabet til forskning i arbejderbevægelsens historie: Copenhagen.

Tanner, Carolyn and Swinbank, Alan (1987), 'Prospects for reform of the Common Agricultural Policy', *Food Policy*, vol. 12, no. 4, November, pp. 290-4.

TCO (1987), *En förändrad livsmedelspolitik*, Stockholm.

Thomsen, Jens Peter Frølund (1996), *British Politics and Trade Union Pressure in 1980s: Governing Against Pressure*, Dartmouth: Aldershot.

Thomsen, Søren Risbjerg (1987), *Danish Elections 1920-79: A Logic Approach to Ecological Analysis and Interference*, Politica: Aarhus.

Thullberg, Per (1974), 'SAP och Jordbruksnärngen 1920-1940: Från klassekamp til folkhem', in *Arbetarrörelsens årsbok 1974*, Bokförlaget Prisma: Stockholm.

Thullberg, Per (1977), *Bönder går samman: En Studie i Riksförbundet Landsbygdens Folk under världkrisen 1929-1933*, LTs förlag: Stockholm.

Tracy, Michael (1989), *Government and Agriculture in Western Europe 1880-1988*, Harvester Wheatsheaf: London.

Tracy, Michael (1993), *Food and Agriculture in a Market Economy: An Introduction to Theory, Practice and Policy*, Agricultural Policy Studies: La Hutte.

Truman, David B. (1951), *The Governmental Process: Political Interests and the Public Opinion*, Alfred A. Knopf: New York.

Tsebelis, George (1990), *Nested Games: Rational Choice in Comparative Politics*, University of California Press: Berkeley.

Udenrigsministeriet (1986), *Beretning fra den danske delegation til GATT-ministermødet i Punta del Este, Uruguay 1986*, Copenhagen.

Udvalget vedr. en Bæredygtig Landbrugsudvikling (1991), 'Beretning om en bæredygtig landbrugsudvikling (miljøinitiativer)', *Folketingstidende 1991-92*, tillæg B, Copenhagen.

Vail, David, Hasund, Knut Per and Drake, Lars (1994), *The Greening of Agricultural Policy in Industrial Societies: Swedish Reforms in Comparative Perspective*, Cornell University Press: Ithaca.

Van Waarden, Franz (1992). 'Dimensions and Types of Policy Networks', *European Journal of Political Research*, vol. 21, no. 1-2, pp. 29-52.

Vedung, Evert (1991), *Utvärdering i politik och förvaltning*, Studentlitteratur: Lund.

Vedung, Evert (1997), 'Policy Instruments: Typologies and Theories', in Marie-Louise Bemalmans-Videc, Ray C. Rist and Evert Vedung (eds), *Policy Instruments and Evaluation*, Transaction Books: New Brunswick, (forthcoming).

Wilks, Stephen (1989), 'Government-industry Relations: Progress and Findings of the ESRC Research Initiative', *Public Administration*, vol. 67, pp. 329-39.

Wilks, Stephen and Wright, Maurice (1987), 'Conclusion: Comparing Government-Industry Relations: States, Sectors, and Networks', in Stephen Wilks and Maurice Wright (eds), *Comparing Government-Industry Relations: Western Europe, the United States, and Japan*, Clarendon Press: Oxford, pp. 274-313.

Wilks, Stephen and Wright, Maurice (eds) (1987a), *Comparing Government-Industry Relations: Western Europe, the United States, and Japan*, Clarendon Press: Oxford.

Williamson, Peter J. (1989), *Corporatism in Perspective: An Introductory Guide to Corporatist Theory*, Sage: London.

Wilson, Frank L. (1983), 'Interest Groups and Politics in Western Europe: The Neo-Corporatist Approach', *Comparative Politics*, vol. 16, no. 1, pp. 105-23.

Wilson, Graham (1990), *Interest Groups*, Blackwell Publishers: Oxford.

Wilson, James Q. (1980), 'The Politics of Regulation', in James Q. Wilson (ed.), *The Politics of Regulation*, Basic Books: New York, pp. 357-94.

Windhoff-Héritier, Adrienne (1991), 'Institutions, Interests and Political Choice' in Roland Czada and Adrienne Windhoff-Héritier (eds), *Political Choice: Institutions, Rules, and the Limits of Rationality,* Campus Verlag/Westview Press: Frankfurt am Main/Boulder, pp. 27-51.

Winter, Søren (1994), *Implementering og effektivitet,* Systime: Herning.

Wistow, Gerald (1992), 'The Health Service Policy Community: Professionals Preeminent or Under Challenge?', in David Marsh and R.A.W. Rhodes (eds), *Policy Networks in British Government,* Clarendon Press: Oxford, pp. 51-74.

Wolman, Harold (1995), 'Policy Networks and Public Policy: Finding a Way Out of the Dead End', Department of Political Science and College of Urban, Labor and Metropolitan Affairs, Wayne State University, Detroit.

Wright, Maurice (1988), 'Policy Community, Policy Network and Comparative Industrial Policies', *Political Studies,* vol. 36, no. 4, pp. 593-612.

Yin, Robert K. (1989), *Case Study Research: Design and Methods,* Sage: London.

Öberg, PerOla (1994), *Särintresse och allmänintrese: Korporatismens ansikten,* skrifter utgivna av Statsvetenskapliga föreningen i Uppsala, 122, Acta Universitatis Upsaliensis, Statsvetenskapliga föreningen: Uppsala.

Index

non-point nitrate pollution, 95
non-structural power, 45-46; 50; 52; 56-57
NPO report, 86-88; 90

O

objective method, 163; 165
operational definition, 13-14; 69; 74-76; 80-81; 83
outcomes of networks, 40
outsider, 2-5; 7-9; 22; 24; 28; 37; 39-41; 46; 49-53; 146; 185; 189-190

P

parliament, 1; 4; 9-10; 17; 21; 42; 49; 55-61; 63; 65-66; 69; 73-74; 81; 84-85; 97; 99; 135; 137; 139; 156; 172-174; 176- 179; 181; 183-184; 186-187; 190
parliamentary support, 4; 9; 10; 57-58; 61; 73
parliamentary working group, 131; 135; 137-139; 140
plural elitist, 19
pluralism, 18-20; 23; 37; 45; 66
point source pollution, 88; 97; 101
policy change, 2-4; 8-9; 17; 32; 39-40; 46; 48-53; 59; 61-62; 64-65; 68; 70; 72-73; 80; 110; 133; 138; 169; 173; 183; 186-187; 189
policy community, 3; 23-27; 29-30; 32-33; 35; 38-39; 41-44; 46-48; 51; 55; 57-58; 65; 146; 155; 161; 169; 183
policy instrument, 69-72; 74-77; 79-80; 83; 87-94; 96; 99-106; 110; 118; 123; 125-126; 128; 133; 135; 138; 141; 185; 191
policy network, 2-4; 8; 20-22; 30-33; 36; 38-39; 44; 46; 56-57; 59; 64; 110; 168; 188-191
policy objective, 7; 69-72; 83; 89; 91; 93; 101; 103; 107; 110; 127; 136; 138-139; 142; 147; 153; 159; 160; 168-170; 185- 187
policy preferences, 47-48; 51-52; 83; 88; 96; 106; 110; 146; 157; 169; 189
policy principle, 4; 8; 13; 26; 39; 43-44;

47-52; 58; 61; 68-73; 83; 110-111; 113; 138; 142; 146-147; 154-155; 169; 186; 189
political party, 21; 55; 56; 57; 58; 59; 91; 135; 139; 151; 157; 173-174
politicisation, 24; 49; 91; 94-95
polluter pays principle, 5; 7; 39; 71; 83; 89; 107; 113; 115; 145; 172; 185
positive economic instrument, 77
price support, 111-112; 121; 131; 136; 142
principle of state responsibility, 5; 7; 39-40; 71; 83; 107; 117; 120; 138; 142; 145-147; 153-155; 160; 172; 185-186; 190
procedures to approach policy problems, 43; 47; 49; 51; 146; 189
public choice approach, 188
public documents, 15

R

Radical Liberal Party, 89; 92; 173-176; 187
radical policy change, 9; 61; 169; 189
regulatory policy instrument, 75; 77-78; 79; 93-95; 102-106
reliability, 13-15
resource exchange, 28; 30; 47; 49; 166
resource interdependency, 21; 30; 34; 47; 50
Rhodes model, 29-30
Riksdag, 84; 99-100; 102-103; 106; 129-130; 132; 135; 137; 139; 141; 156; 157-158; 177-179; 183

S

second order reform, 73; 132; 186
set aside, 7; 114-115; 120-121; 124; 126- 127; 129
Single European Act, 5; 182
Smallholders' Union, 84; 86; 89; 122; 125; 151
Social Democratic party, 57; 98-99; 157; 159; 175-179
social group, 4; 9; 16; 32; 52; 55; 57-61; 65-66; 172-173; 177; 190
social structure, 54; 59

Printed and bound by CPI Group (UK) Ltd, Croydon, CR0 4YY

21/10/2024

01777087-0006